The Reporter's Handbook on Nuclear Materials, Energy, and Waste Management

The Reporter's Handbook

on Nuclear Materials, Energy, and Waste Management

Michael R. Greenberg
Bernadette M. West
Karen W. Lowrie
Henry J. Mayer

Vanderbilt University Press

Nashville

Nashville, Tennessee 37235

13 12 11 10 09 1 2 3 4 5

This book is printed on acid-free paper made
from 30% post-consumer recycled content.
Manufactured in the United States of America

Library of Congress Cataloging-in-Publication Data
The reporter's handbook on nuclear materials, energy, and
waste management / by Michael R. Greenberg . . . [et al.].
p. cm.
Includes bibliographical references and index.
ISBN 978-0-8265-1659-6 (cloth : alk. paper)
ISBN 978-0-8265-1660-2 (pbk. : alk. paper)
1. Nuclear energy—Press coverage. 2. Nuclear industry—
Press coverage. 3. Radioactive waste disposal—Press
coverage. 4. Spent reactor fuels—Press coverage.
5. Nuclear facilities—Environmental aspects—Press
coverage. 6. Journalism, Scientific.
I. Greenberg, Michael R.
TK9145.R42 2009
363.17'99—dc22
2008038600

Contents

Preface

Journalists, who face the challenge of writing stories that are accurate, balanced, objective, and responsible, have reported about the benefits and risks associated with radionuclides from the days of the Manhattan Project when information was a guarded secret and few knew what was happening to today when a plethora of information and opinions exists. They have written about the beneficial use of radioactivity to kill rapidly growing cancer cells and identify malfunctioning organs and of x-rays to detect caries and other dental problems, the development of devices to accurately measure the thickness and quality of products, the use of radionuclides to kill pathogens, the installation of radionuclide-containing smoke detectors, and many other uses of radioactive materials to improve quality of life and create economic opportunities.

Yet, journalists instantly recognize the words Chernobyl and Three Mile Island, along with acid rain, Bhopal, dioxin, Love Canal, Exxon Valdez, global warming, and ozone depletion as among the environmental stories that rose to front-page and nightly network news headlines. There is nothing simple about preparing an accurate, balanced, objective, and responsible story about radiation and radioactivity.

World events require stories that tie together nuclear power, nuclear waste, nuclear weapons, global warming, economic development, and public health. This need for high-quality reporting about nuclear issues comes at a time when newspapers, radio and television stations, and magazines are under substantial financial pressure. During the 1980s, many media outlets added environmental beat reporters. Most of these specialist jobs have disappeared. Nonspecialist reporters assigned to a breaking or background story probably have relatively little knowledge about radionuclides. We believe they would benefit from a handbook that provides basic information and leads for further research.

Our goal is to provide that handbook; it most certainly is not to try to persuade journalists that radioactivity and its uses are good or bad. In 1988 we published the *Environmental Reporter's Handbook,* using a formula suggested to us by journalists. The book was praised by reviewers and received a special award for journalism from the Sigma Delta Chi Society of professional journalists in 1989. In 1995, we published a second edition, adding more information and changing the title to *The Reporter's Environmental Handbook,* because we learned that most of the users were not environmental reporters but nonspecialist reporters who were covering an environmental story. In 2003, we published a third edition of the handbook. That edition was informed by a survey of the members of the Society of Environmental Journalists, who identified topics they wanted us to cover and helped us tweak the handbook format. The current book is more specialized, focusing on nuclear materials, nuclear energy, and nuclear waste; otherwise it resembles its predecessors.

Part I begins with suggestions about how to use this book most effectively and continues with an introductory essay on why nuclear-related developments have become a major policy issue. The next essay presents synopses of various crosscutting themes, such as environmental impact, risk assessment, and economic analyses, and describes the frameworks used by analysts to assess and weigh the advantages and disadvantages of policy options involving these themes. Part I ends with an essay by Tom Henry, an award-winning journalist with more than 26 years of experience, who writes about how he would cover some of the issues presented in this handbook. Readers will find his brief useful when covering nuclear power; however, he does not focus directly on issues related to waste management and transportation, decommissioning, the economics of nuclear power, weapons, or issues of nonproliferation. These topics are covered elsewhere in this volume.

Part II consists of essays that focus on nuclear-related issues. Journalists indicated that they do not want a science textbook in which they have to thumb through 10,000-word essays with 100 citations of sources that are mostly available only in paper copies in a library. Our briefs of 2,500–4,000 words capture the essence of an issue, such as dirty bombs or nuclear reactor safety. They are the heart of the book. Each brief

- describes the broad background of the issue;
- identifies key questions and issues for journalists to ask in their investigation;
- discusses hazards, risks, and benefits to the public;
- reviews what experts believe are myths and misunderstandings among the public;
- suggests pitfalls that are commonly found in media coverage on the topic; and
- offers resources for follow-up research

We recognize that a 2,500–4,000 word essay on medical uses of radionuclides, dirty bombs, or engineering of nuclear reactors will not satisfy the reporter who specializes in the subject. That reporter will find the information in the briefs to be too basic. Rather, our target, as noted earlier, is the reporter who does not have much of a background with nuclear topics and will probably be covering one or two other stories, perhaps about crime, politics, and health, at the same time. He or she may need a reliable concise source of information as a starting point that can be read in 20 minutes and offers readily accessible sources for follow-up. The briefs in this handbook should serve that purpose. Several of the briefs are longer than 4,000 words because our experts told us that the initial draft did not convey sufficient basic information.

Journalists read background materials but also rely on expert sources for information. For this book, as for its predecessors, we interviewed leading experts from universities, business, government, and citizens groups. Some of the briefs will appear to be slanted in one direction or the other because experts, like everyone else, have viewpoints. Yet, it was critical for us that the book be as balanced as possible. Consequently, every brief in the book has been reviewed by an external panel of individuals who, while they may have different viewpoints, have expertise on this subject.

Part III includes a glossary, a summary of key laws and policies, a summary of the history of nuclear power, and a list of organizations, with a description of each one, that may be consulted for detailed reviews of the subjects covered in this book. For example, the American Nuclear Society, founded in 1954, is a not-for-profit international organization consisting of over 10,000 engineers, scientists, professors, students, and others interested in nuclear issues. The vast majority of members live outside the United States. Notably, the society has issued "position statements" on almost 40 subjects, including nonproliferation, health effects of low-level radiation, disposition of surplus weapons plutonium, transporting nuclear waste, and many others.

Several notes are in order about the topics covered in this volume. Early on, we recognized that though it was not feasible to cover non-nuclear topics in depth, reporters need quick access to information about related topics. Thus, for example, there is a single brief about climate change and non-nuclear energy options that summarizes the issue and its relationships to fossil fuels, renewable sources, and conservation.

It was also not feasible to provide equal coverage about every nuclear-related topic. Within the nuclear topics, we emphasize nuclear waste management and nuclear power and devote less attention to military uses of nuclear materials, such as nuclear weapons and nuclear-powered submarines and other U.S. Navy ships. Other choices about what to include followed from conversations with journalists. They asked us to look back at what has happened as a result of the Three Mile Island and Chernobyl events, and they asked us to prepare a brief about the protection of current nuclear power plants against terrorists. Less attention was devoted to licensing of existing nuclear power plants. Also in response to these conversations, we emphasized the science and engineering issues rather than economics, politics, and communication issues, which are included in single briefs. Finally, our geographical focus is the United States, although international issues are found throughout the volume.

Avoiding actual or even perceived bias in the handbook has been our major concern. To ensure minimum bias, we isolated the authors from the review panel. The experts who reviewed the briefs were not chosen by the authors. All correspondence and review of materials was done through the Consortium for Risk Evaluation with Stakeholder Participation (CRESP), a multi-university group funded by the U.S. Department of Energy (DOE) to conduct research. Also, in response to a concern that readers would not encounter a range of responsible views and criticisms of the topics presented in this volume because of the potential for bias in the authors' choices of sources, we added a section to the resources in Part III called "Key Sources." Here we list classic and recent books and articles, and Web sites for organizations and individuals that support and oppose nuclear power, reprocessing, and some other applications of nuclear technology.

In addition to consulting the specialized sources throughout this volume, reporters should look to their state and local health departments for information. Almost every one has an individual or group that deals with radiation health and is involved with preparedness. We are not saying that these institutions replace national experts, we are saying that for local stories they must be consulted.

The authors are greatly indebted to many individuals who participated in the preparation of this volume. First, we thank the scientists who were interviewed for the briefs. These are listed on the briefs, and a short biography is provided for each of them.

We are a deeply grateful to CRESP for supporting the project and for organizing and conducting the peer review. We thank Charles Powers and David Kosson for their encouragement and patience. We thank Arthur Upton and Bernard Goldstein for chairing the peer review panel and for organizing the peer review of the entire document. We also thank Milton Russell and Don Hopey who reviewed the entire handbook. Short biographies of the reviewers are found at the end of the volume.

The Reporter's Handbook on Nuclear Materials, Energy, and Waste Management is based on work supported by the DOE, under Cooperative Agreement Number DE-FC01-06EW07053 entitled "The Consortium for Risk Evaluation with Stakeholder Participation III" awarded to Vanderbilt University. The opinions, findings, conclusions, or recommendations expressed herein are those of the authors and do not necessarily represent the views of the DOE or Vanderbilt University. Obviously, the DOE has a stake in nuclear-related issues. While the DOE funded the handbook, it did not instruct the authors about the topics to be included, and it did not exercise editorial control over the handbook. Four of the 21 briefs are partly based on interviews with DOE subject-matter experts. Also, as noted, all briefs were peer reviewed by our expert panel for accuracy and objectivity.

A closing note: the media too frequently are blamed for bad public policy decisions, and the media probably do not receive enough credit for good policy decisions. But no one denies the importance of their efforts to communicate accurate, balanced, objective, and responsible stories.

Disclaimer: This report was prepared as an account of work sponsored by an Agency of the United States Government. Neither the United States Government nor any agency thereof, nor any of their employees, makes any warranty, express or implied, or assumes any legal liability or responsibility for the accuracy, completeness, or usefulness of any information, apparatus, product, or process disclosed, or represents that its use would not infringe privately owned rights. Reference herein to any specific commercial product, process, or service by trade name, trademark, manufacturer, or otherwise does not necessarily constitute or imply its endorsement, recommendation, or favoring by the United States Government or any agency thereof.

About CRESP

Consortium for Risk Evaluation with Stakeholder Participation III

The Consortium for Risk Evaluation and Stakeholder Participation (CRESP), since 1995, has been researching ways to advance cost-effective cleanup of the nation's nuclear weapons production waste sites and test facilities. The consortium responded to a request by the U.S. Department of Energy (DOE) and the National Research Council for the creation of an independent institutional mechanism to develop data and methodology to make risk and stakeholder involvement a key part of decision making at the Environmental Management Office of DOE. As a result of a national competition, a 5-year cooperative agreement was awarded to CRESP in March 1995. The CRESP co-founders and initial management board included Bernard D. Goldstein, John A. Moore, Gilbert S. Omenn, Charles W. Powers, and Arthur C. Upton. CRESP I was institutionally managed by the Environmental Occupational Health Sciences Institute in New Jersey with Bernard D. Goldstein as principal investigator and Charles W. Powers as executive director.

The first 5-year cooperative agreement was renewed in 2000 with the Institute for Responsible Management as the lead institution and Charles W. Powers as principal investigator. CRESP worked to improve the scientific and technical basis of environmental management decisions leading to advance protective and cost-effective cleanup of the nation's nuclear weapons, and to enhance stakeholder understanding of the nation's nuclear weapons production facility waste sites. CRESP II pursued this work through a unique institutional model: (1) its primary mode of operation was an unprecedented program of interdisciplinary, multi-university research; (2) it was independent and its beneficiaries are those who have a stake in effective cleanup of federal facilities; and (3) it was organized to provide guidance to and peer review of the evolving effort to use risk methods and evaluations to shape cleanup decisions at DOE sites. All three elements were effectively demonstrated in CRESP's work on key problems at both major and "small" DOE Environmental Management sites, especially Amchitka, a volcanic island that is part of the Aleutian islands of Alaska.

CRESP III was renewed as a DOE cooperative agreement in the fall of 2006 with Vanderbilt University as the lead organization, and Charles W. Powers and David S. Kosson as co-principal investigators. The objective of the CRESP III project is to advance cost-effective, risk-informed cleanup of the nation's nuclear weapons production facility waste sites and cost-effective, risk-informed management of potential future nuclear sites and wastes. This objective is being accomplished by seeking to improve the scientific and technical basis for environmental management decisions by the DOE, and by fostering public participation in that search. The CRESP III member colleges and universities now include Howard University, New York University School of Law, Oregon State University, Robert Wood Johnson Medical School, Rutgers, the State University of New Jersey, University of Arizona, the University of Pittsburgh, and the University of Washington.

Part I: Getting Started

How to Use the Handbook

If you need just a definition or quick explanation, go directly to the glossary in Part III. For example, if you want to know what a "curie" is, go to the glossary and look up the definition. You will find that it is "the basic unit used to describe the intensity of radioactivity in a sample of material. The curie is equal to 37 billion (3.7×10^{10}) disintegrations per second, which is approximately the activity of 1 gram of radium." If you want a more detailed discussion of a particular topic, go to the table of contents and look over the topics in Part II. For example, for information on how a curie relates to a becquerel and how both are related to health impacts, turn to Section 1 of Part II. It provides background information about radionuclides, key issues related to public health, possible stories, and pitfalls noted in previous coverage of health impacts.

Each brief is self-contained. Reporters should be able to get what they need from a given brief; in other words, we hoped to reduce the need to search through the book. This means that there is some unavoidable redundancy among the briefs. To assist those who need more information, within the briefs, we cross-reference other briefs. The index provides further guidance.

We clustered the briefs into five sections within Part II. Section 1 examines nuclear materials and radioactivity—that is, what they are, how they are formed, where they are found, and most important, effects of radiation on humans. Section 2 examines nuclear power and other nonmilitary uses of radionuclides, including nuclear medicine and food irradiation. It explores issues that have arisen during the past half-century, such as nuclear-energy safety systems, the Chernobyl and Three Mile Island events, and the economics of nuclear power. Section 3 focuses on nuclear waste management. Briefs describe nuclear waste, how and where it is managed, monitoring of waste management sites, the ecological impacts of cleanup, and long-term surveillance and maintenance of waste management sites. Section 4 focuses on military-related nuclear issues, such as managing nuclear weapons, radiological dispersal devices (dirty bombs), nonproliferation initiatives, nuclear terrorism, and international and national policy related to these. Section 5 reviews climate change, public perception, and risk communication focused on nuclear energy and waste issues.

Reporters who are not familiar with environmental risk, economic, and technology assessment, risk perception, and theories about how technology fits into the larger context of resource management will find helpful background in the short overviews in Part I, "Crosscutting Themes."

Why Now? Why This Discussion?

Written by Michael R. Greenberg, with comments by
John F. Ahearne and Richard L. Garwin

The simple answer to "Why now?" is that the governments and people of the world are being driven to consider nuclear power and other energy sources, along with conservation, as options for meeting increasing energy demand. This is not the first time this pressure has gripped the United States, but the increasing fear about climate change has added another dimension. Also, the United States, Russia, France, and Great Britain face major nuclear weapons waste issues as a cold war legacy.

Beginning with the nuclear energy issue, on October 17, 1973, the members of the OAPEC (Organization of Arab Petroleum Exporting Countries) embargoed petroleum shipments to the United States, some of Israel's allies in Western Europe (initially the Netherlands) and Japan because of their support for Israel against Egypt and Syria in the Yom Kippur War. Just before the oil embargo in 1973, the average gas price at the pump was $1.80 per gallon (adjusted for inflation to 2007 dollars). In 1981, the average price was $3.00 (a 70% increase). These price increases sent a recessionary ripple through the economies of the dependent nations that spread across the world. High oil prices persisted until 1986. The embargo and price increases sparked an interest in exploration for conservation and new sources of fossil fuels. Governments' monetary policies became more restrictive, and interest in nuclear power increased.

France, Belgium, Sweden, and Japan now heavily depend on nuclear power. In the United States, even before the Three Mile Island nuclear reactor meltdown in 1979, U.S. commercial business interest in nuclear power was waning. A worldwide recession during the oil embargo caused economists to reduce their estimates of the growth of electricity demand, and the price of new reactors seemed high to U.S. utilities. Furthermore, the U.S. economy grew despite the lack of growth of energy use. Serious efforts were made by all sectors of the U.S. economy to economize energy use. After 1986, the year of the Chernobyl nuclear incident, the economy continued to grow; while oil prices declined and remained relatively low until the new millennium. Reprocessing nuclear fuel that has been used once in a nuclear reactor to generate electricity was considered too risky by the United States because it has the potential to be used for nuclear weapons proliferation. The incident at Chernobyl along with the extensive time required to construct and license nuclear power plants increased costs and further undermined the credibility of nuclear power. National leaders and utilities concluded that nuclear power in the United States was a bad idea. Other countries, such as Japan and France did not agree and moved forward with nuclear power plant operations.

The events of the current decade are forcing reconsideration of policies examined during the embargo and price increases of the 1970s. The political instability of the world's oil producing nations has created a fear of political blackmail by petroleum supplying nations in the United States and other countries. The rapid rise and fall of petroleum prices seems inexplicable even to some experts. Also, a new consideration is that during the past decade scientists have become convinced that the burning of fossil fuels is leading to global warming, whereas nuclear power does not contribute notably to greenhouse gases that lead to global warming. Therefore, nuclear power, despite its history of environmental and economic risks and despite waste management problems, seems to some like an environmental bargain. Proponents of nuclear power argue that France and Japan have successfully invested in nuclear power generation and have not suffered obvious environmental problems. The United States and other Western nations have observed the rapid growth of the Chinese and Indian economies and are concerned that competition for oil and gas will drive up prices still further and thereby undermine the economies of the developed nations. Nuclear power seems like a logical approach. Yet, there is obvious dissent from this position. Some propose a reduction in the use of carbon and nuclear fuel, arguing for eliminating subsidies for carbon-based and nuclear-based fuels, for heavy investment in solar and other renewable technologies, and for other policy changes that would largely achieve fossil and nuclear fuel reduction objectives in 30 to 50 years (see Makhijani, 2007). Often lost in these discussions are the distinctions between transportation fuels and energy sources for electricity production and industrial use, as well as intermittent (such as from wind) and peak load power production (often from natural gas) and base load power production (such as from coal and nuclear energy). "Used" nuclear fuel, that is, nuclear fuel that has been used once in a nuclear power plant, can be reused as nuclear fuel after conversion; "spent" nuclear fuel no longer has the capacity to be recycled as nuclear fuel. The used/spent nuclear fuel issue is sometimes lost in the very public discussion of nuclear power because the plutonium can be extracted and manufactured into the right shape and joined with high explosives and a trigger device to produce nuclear weapons. Turning spent nuclear fuel into a nuclear weapon, however, is extremely difficult. Reprocessing, manufacturing, and building trigger devices to produce nuclear weapons is an extraordinarily complicated task, and fuel from commercial nuclear plants has not been used by a proliferate state. Although technically once-through nuclear fuel is "used" not "spent," the literature refers to it as spent. For consistency with the literature, we call it spent in this volume.

While not as prominent in the public eye as nuclear power, legacy wastes from the production of nuclear weapons place a major environmental management burden on the U.S. Department of Energy and are the focus of the most costly government environmental management program in the world. Between 1989 and 2007, the Department of Energy spent an estimated $80 billion on managing the waste at over 130 sites across the United States. Many of the sites have closed, but the challenge of controlling high-level waste at the Hanford (WA), Idaho National Laboratory (ID), Oak Ridge (TN), and Savannah River (SC) sites remains. These military waste sites are so technologically, environmentally, legally, and economically challenging that they will need management in perpetuity. While legally the civilian nuclear waste stream and the defense-related waste stream are managed separately, it can be argued that they intersect and that methods used to manage one stream can be applied to the other. There are political and economic reasons to separate and other arguments to combine the civilian and military nuclear waste streams. Russia, with many nuclear weapons and much waste, faces a similar challenge.

In short, the stakes for the world's nations in considering these issues now have never been higher. The media as always are expected to meet the challenge of writing stories that are accurate, balanced, objective, and responsible about the individual elements and the overall puzzle that ties together nuclear power, nuclear waste, nuclear weapons, global warming, economic development, and public health. It is our hope that this volume will contribute information to what most certainly will be a tough debate.

Reference

Makhijani, A. (2007, August). Carbon-free and nuclear-free: A roadmap for U.S. energy policy. *Science for Democratic Action,* 15(1).

Crosscutting Themes

Written by Michael R. Greenberg, with comments by
John F. Ahearne and Richard L. Garwin

Five themes are central to the issues discussed in this handbook and are explicit or implicit in every brief. These themes are as follows: (1) environmental impact, (2) risk, (3) economics, (4) evidence and public perception, and (5) ripple effects of decisions. Each of these themes is a massive subject by itself. We make no pretense of providing a comprehensive review. The goal is rather to draw the reader's attention to how these themes are core to nuclear power, waste management, and other nuclear-related issues.

This handbook presents the views of leading experts in nuclear-related issues that during the next decade will require enhanced media coverage. These experts hold strong opinions and reflect a diversity of viewpoints, and it would be wrong to imply that this book presents or indirectly implies a unified viewpoint about nuclear-related issues. It does not. We were not interested in capturing their opinions; rather we tried to understand and then summarize their very nuanced understanding of important parts of this complex subject. In the course of these conversations, however, and as the review process unfolded, we recognized that while these experts might strongly disagree with each other about the details of nuclear-related issues, they all have a deep respect for and concern about the challenges of resource management and technology in a rapidly growing world economy. Behind the details of each brief is the larger story of global resource management and technology, the subject of the last part of this section. Perhaps, it is the big story within the many smaller stories to be written.

Environmental Impact

On January 1, 1970, the National Environmental Policy Act (NEPA) became national policy. Its goal was to "create and maintain conditions under which man and nature can exist in productive harmony, and fulfill the social, economic, and other requirements of present and future generations of Americans" (P.L. 91-190, Chapter 55, 4331(9)). Behind the impressive objective, the law led to the requirement that an environmental impact assessment would be prepared to ensure that major federal government projects or programs would undergo comprehensive review before construction or implementation. The review entails a multidisciplinary, multi-agency public assessment of the environmental, economic, health, and social impacts of individual projects or program proposals, as well as the consideration of alternatives.

Environmental analyses for nuclear-related programs and individual projects are multi-thousand-page documents costing millions of dollars. The U.S. Department of Energy's massive nuclear waste management program has had numerous programmatic and project-specific assessments. For example, the Yucca Mountain Organization in Eureka County, Nevada, maintains a Web site that lists and discusses the environmental impact statement (EIS) work on the proposed final repository (*www.yuccamountain.org/eis.htm*). Utilities that want to build a nuclear power plant are required to prepare an environmental assessment. For example, the first author and a colleague prepared the population, land use, and economic elements of the EIS for a proposed nuclear power plant in the Delaware River. That Newbold Island EIS was thousands of pages long. The license was not granted to the utility, in part because of the finding of excess population density in the immediate vicinity of the proposed nuclear power plant.

Whether 200 or 2,000 pages, each EIS is required to contain a detailed description of the following five factors:

1. The environmental impact of the proposed action
2. Any adverse environmental effects that cannot be avoided if the proposed action is implemented
3. Alternatives to the proposed action
4. The relationship between local short-term uses of the human environment and the maintenance and enhancement of long-term productivity
5. Any irreversible and irretrievable commitments of resources that would be involved if the proposed action is implemented

In the original act, states and municipalities "owe no duties under NEPA, but may be subject to alternative environmental legislation fashioned after NEPA." In fact, many states and local governments have NEPA progeny that require private interests to prepare environmental assessments.

Although the wording varies by agency, the following six topics encompass the essence of environmental impact requirements:

1. Description of the existing environment
2. Description of alternatives
3. Probable impacts of each alternative
4. Identification of the alternative chosen and the evaluation that led to this choice
5. Detailed analysis of the probable impacts of the proposal
6. Description of the techniques intended to minimize any adverse impacts

In addition to generating a good deal of information that has stopped or modified project decisions, the EIS process creates checks and balances among federal, state, and local governments. A federal department may ignore opposition by other government agencies to a proposed program, but this usually does not happen. Public participation can be enhanced because the EIS requires the agency to read and respond to all comments. In short, the EIS process should lead to more effective thinking and planning, raise agency and general public awareness, and create a series of checks and balances.

NEPA has led to the cancellation or postponement of proposals to build dams, airports, highways, nuclear waste disposal programs, outer continental shelf leases, and other projects. More frequently, the process has resulted in design changes, location changes, and other modifications. Yet, critics have not been satisfied with the process. The key issue is that agencies are not obligated to change their decisions. Also, some agencies have decided that their projects are not "major" or "significant" and do not constitute an agency "proposal" or "action." Finally, length and detail do not necessarily mean that all the scientifically ascertainable impacts are included. Some scientific facts will be missing because information is lacking.

Summarizing, every major nuclear power and defense program initiated since 1970 has required the preparation of environmental analyses that have led to changes, some major and others minor.

Risk

Risk analysis is a multistage process for determining the likelihood of adverse human and ecological effects of exposure to biological, chemical, and physical hazards and then reducing the risk. Potential hazards include toxins (e.g., asbestos), structures (e.g., dams), and activities (e.g., driving while intoxicated). We briefly describe the process, highlighting some of the key strengths and weaknesses and focusing on hazardous materials. The seven steps in the process would be slightly altered for structures and activities. Risk depends on (1) the hazardousness of the material, (2) its quantity, (3) the probability of release, (4) the dispersion of the hazard, (5) the population exposed, (6) organism uptake, and (7) response of officials to the hazard before, during, and after release.

To assess the hazardousness of a material, scientists have identified hazards by studying historical records of human exposure and health outcomes. Research on asbestos is an excellent example of how retrospective data is used to link a strong hazard and diseases. Sometimes researchers follow people for years and observe as some become ill and others do not. The Framingham doctors and nurses studies are three multiyear large-sample prospective studies that have produced important findings about the hazards of high cholesterol, smoking, and other hazards, and the advantages of behaviors such as exercising. The bulk of assessments, though, rely on laboratory testing of mice and other species that are sentinels for human effects, which does not always work to identify human hazards.

After gathering data, scientists use statistical models to estimate the excess risk associated with exposure to the likely hazard. Some models assume that human response increases directly with dose, that the relationship is linear. Others—called threshold models—assume that some substances are harmless unless a threshold dose is reached. There is also a set of nonlinear threshold models and a set of no-threshold models. Debates are common because the form of the model adopted by government regulators influences the risk estimates. Since adequate low-dose exposure and response data are difficult and expensive to acquire and thus are scarce, policy makers must decide which model assumption to use based on higher exposure data where the effects are more frequent and then must make assumptions about the effects at low doses. With regard to radioactivity, the linear assumption has prevailed as a conservative model.

Preventing a hazardous substance from being released is a key component of risk assessment and management. If we know the hazard posed by a substance and we know its location, then we should be able to contain it. But managers often resist spending money on containment if a problem is unlikely to occur, because there are always other pressing needs for limited resources.

Dispersion of hazards is another concern in risk analysis. Contaminants can spread through direct contact, air, water, and soil. Scientists have developed mathematical and physical models to detect the spread of hazards. For example, water quality dispersion models are widely used in risk analyses to estimate the impact of discharges into rivers from electricity-generating stations, refineries, and sewage plants.

The human and ecological population at risk is also considered. Some hazards are ubiquitous, such as cigarette smoke and auto emissions. Others are not. Atomic Energy Commission (now U.S. Nuclear Regulatory Commission) guidelines require that nuclear power plants not be located in urban centers. Some nuclear plants have not been built or never operated as a result of risk estimates, for example, one proposed near Trenton, New Jersey, and Philadelphia, another proposed in New York City directly across the river from the United Nations headquarters, and a third that was built in Shoreham, New York, but never operated. The first two were relocated to more "remote" locations. However, one of these, Indian Point, New York, is no longer a remote location and has been a focal point for criticism about the location of nuclear power plants. One of our reviewers pointed out that U.S. policies regarding dispersion studies are not consistent. For example, he observed that radiation dispersion studies are not required for coal plants even though the amount of radioactivity they emit equals or exceeds that from a nuclear power plant (from radioactive elements naturally present in coal). Likewise, the requirements for public and private facilities that emit pollutants may be different.

The final components of risk assessment and management are uptake of the hazard and response to the dose, and government, private organization, and personal response to a hazard. Some people and species are more sensitive to some hazards than are others. Government agencies normally assume those exposed are the most or among the most sensitive people when they set environmental exposure standards. For example, the ambient air quality standard for lead is set for children, whose central nervous systems are sensitive to lead uptake.

Using accurate and rapid risk characterization and communication, government and individuals can prevent or reduce exposure or the effects of exposure. For example, all nuclear power plants are required to have a buffer area around them in which there are no homes. Power plant managers can devise other protective systems to reduce exposure, including shielding. For example, alpha particle penetration can be stopped by a single sheet of paper, whereas beta particles, which have a much smaller mass and much higher speeds, require more material, or denser material, such as lead or iron to stop penetration. Should a notable emission occur, sheltered protection and sometimes evacuation may be required. In some cases, prophylactic steps may be possible. For example, a person can consume stable iodine to prevent radioactive iodine 125 and 131 from being absorbed by the thyroid.

Risk analysis has its detractors and proponents. A prominent criticism is that risk estimates are sometimes too uncertain. In some cases, the high estimate of risk is double the low estimate, and sometimes the high estimate is greater than 10 times more than the low estimate. When the range is so great, the results may not be helpful to managers who must make a yes or no decision. Some critics argue that risk analysis is a way of hiding ethical choices in a body of undecipherable numbers and unstated assumptions, and that the disadvantaged end up bearing a disproportionate amount of the risk burden. Proponents of risk analysis argue that an orderly presentation of data, including disclosure about uncertainty, provides managers with important information and alerts researchers to shortcomings that can be addressed. Further, they would assert that many of the criticisms of risk assessments are being addressed.

In 1990 the federal government created the Commission on Risk Assessment and Risk Management to recommend how risk analysis could be most effectively used in a regulatory framework. The commission published several volumes and seven appendices, all of which are available on the Internet (*www.riskworld.com/riskcommission/default.html*).

Overall, during the past 2 decades, risk analysis has become a key tool used by engineers and scientists to assess and modify technologies and operations in the nuclear power industry, and in the military and civilian waste management industries.

Economics

To measure the total cost of a product, process, or facility, engineers and economists use life-cycle cost analysis (LCCA). LCCA starts with a concept, design, plan, and development. The second phase, usually the most expensive, includes obtaining land, building facilities, and installing the processes, then operating and maintaining them. The third phase involves disposal, which for a plant may mean all or part of it must be rebuilt or closed down. Some facilities will require remediation and long-term stewardship.

One of the most complex uses of LCCA is in high-level nuclear waste management because of the long half-life of some nuclear elements. A precise estimate of the cost of maintaining facilities that may be needed for thousands of years is beyond our current capacity. The best we can reasonably expect is plausible multigenerational estimates.

In addition to the three phases of LCCA, economists, planners, and engineers must take into account discount rates and depreciation. The discount rate is the interest rate charged for a loan. For federal government projects, the Office of Management and Budget provides guidelines on the value of money and the cost of borrowing it. Depreciation is the decrease in economic value because of obsolescence, physical deterioration, and losses in the utility of a facility during its productive life.

LCCA has advantages and disadvantages. One advantage is that all costs become transparent, that is, decision makers can see an estimate of what they would be spending in the short term and in the long term. Another advantage is that trade-offs among technologies and operations become clear. For example, a cheaper overall life-cycle cost may assume technological innovations that may not occur. Decision makers can probe those assumptions and decide whether they are too risky.

The longer the expected life-span of the activity and facilities, the more uncertain the cost assumptions and estimates. With regard to nuclear power in the United States, the utility must maintain, possibly replace parts of, and eventually decontaminate and decommission large facilities. In addition, the federal government and utilities must design, build, and maintain facilities to store and manage hazardous materials for hundreds to thousands of years.

Much of the literature about LCCA is written for engineers and economists. For example, major reports and books have been written about LCCA for bridges, roads and pavements, air traffic control, and other key pieces of infrastructure. The U.S. Department of Energy and the U.S. Nuclear Regulatory Commission face an unprecedented challenge in estimating the cost of facilities that are expected to last hundreds to thousands of years.

If "How much will it cost?" is the first question, then "How much will it benefit and who will benefit?" is the second question. Economists, regional scientists, and geographers have developed a set of economic analysis tools that can help answer the economic impact policy questions. The most widely used models use historical records and economic trends from 25 to 30 years back to prepare equations that estimate future economic impacts.

The strength of these models is in their use of historical trends, but that is also one of their weaknesses. Past economic relationships do not always provide accurate predictions of the future. These models cannot account, for example, for major or sudden economic shifts that drastically change the pattern of historical business transactions. Predictions will be only as good as the data used in the models. With massive engineering projects, like those discussed in this book, we need estimates of economic resiliency that take into account unanticipated bottlenecks or opportunities. When the economy does recover, how will it have changed in ways that impact the programs and projects being analyzed?

Evidence and Public Perception

Meshing the scientist's and the public's perspectives on nuclear power, waste management, and technologies is challenging because we have strong evidence that they use different criteria for weighing evidence and deciding what to believe. Scientists are supposed to leave their preferences out of their work and, by training and practice, consider the five following attributes of evidence that come before them.

1. Rigor. A good study will have clear and answerable research questions, and it should describe data and methods in detail. A rigorous study uses the best scientific practices or explains why they could not be used. It also describes any limitations and their implications.
2. Corroboration. A good study will be confirmed by multiple independent scientists.
3. Power. A good study will have enough samples that the effect it was designed to look for will be easily detected.
4. Universality. A good study will show similar results among different test groups, such as, in a laboratory study, multiple species or several exposure routes (e.g., skin, respiratory system, digestive system). Similarly, in a study involving spatial data, scientists would look for similar results in different geographical locations, among different cultures, or across racial/ethnic and income/education groups.
5. Relevance. A good study will have enough evidence to allow scientists to attribute the effects to appropriate social, economic, political, chemical, physical, biological, or other theoretical constructs.

In contrast, Sandman (1993) maintains that public reaction to hazard and risk is driven by "outrage" factors. He has described more than 20 factors that cause people to evaluate some hazards as higher than others but that do not relate to their scientifically calculated risk. We list nine of them with brief examples to show the remarkably different ways experts and the public evaluate risk.

1. Lack of control. Holding a nail while someone else hammers it into a wall seems significantly more dangerous than doing the hammering oneself.
2. Imposition/coercion. Vaccinations that are required seem riskier than allergy injections one volunteers to receive.
3. Inequity. Locating a county incinerator in a poor neighborhood that did not volunteer to receive it and that already has multiple risks seems more risky

than locating it in a wealthier neighborhood or business district that has few other risks.

4. Lack of familiarity. A new chemical agent seems riskier than tobacco smoke.
5. Memorability. The negative image of a nuclear mushroom cloud is virtually indelible.
6. Concentration. Fear is increased by events in which large numbers of people die or are injured in a short time or in a confined area, such as in the crash of a large passenger aircraft or destruction by a hurricane.
7. Immorality. Actions such as using animals for scientific research or cutting down a forest are rejected because they are considered to be immoral, irrespective of the actual risks they involve.
8. Lack of candor. Learning from the media that a road is being put through one's neighborhood after city officials insisted it was only under preliminary consideration increases the sense of risk associated with the project.
9. Human-made risks. Hazards caused by human error, such as an oil spill, are considered to be more dangerous than natural hazards, such as a flood, even if the actual impacts are similar.

Arguably, these nine outrage factors are a distraction from reality, leading people to make bad choices that are not supported by science. Yet, they have been observed in studies around the world among men and women in different age, ethnic, racial, religious, and socioeconomic groups. These outrage factors are psychologically driven guides that people use consciously and subconsciously to consider evidence.

Reporters must respect both the science and the outrage factors, recognizing that whereas scientists trust authority and expertise, weigh accumulated evidence, rely on scientific theories and consistent findings, and live with uncertainty, the public trusts traditions, peer groups, open processes, and notably widespread public participation, and they will consider anecdotal information and worry about uncertainty. The American public distrusts linear, reductionist, and expert-based decision-making processes funded by those it considers to have a vested interest in the outcome. They want an opportunity to participate and not have the scope limited to discussions of "proven" quantitative scientific results.

Ripple Effects of Decisions

We cannot precisely follow the impacts of decisions to every location, nor can we say with confidence what will be the long-term impacts of an action. Yet, we can try to foresee and try to understand the major effects. Reporters may find that some of the distant ripples are worth investigating.

To bring the metaphor to life, we pose a hypothetical policy question: Should a new electricity peaking facility that burns natural gas be built adjacent to a river on a brownfield site that was abandoned 20 years ago? Alternatively, should the facility be constructed at an existing coal-fueled baseload electricity site 10 miles away? Or should nothing be built locally?

For context, the area on the side of the river opposite the proposed location has only a few houses and the river is used mostly by sport fishermen and boating enthusiasts. Elected officials on this relatively undeveloped side oppose the facility. Their counterparts on the side of the river with a brownfield strongly support it.

The first ripple is the direct environmental impact and the immediate financial impact associated with the proposed new electricity plant. For example, the old factories will be demolished and the site will be remediated. This means that some of the residual contamination will be remediated and the remainder will be left in place and capped. The project will necessitate short-term disruption of the river flow away from the location while the facility is constructed and when pipelines are extended to supply the natural gas to the site. Some fish and other species will find their passage more difficult when the flow of the river is temporarily changed. Small-boat traffic can continue but will be slowed down for a time. Doubtless, there will be some work-related injuries during the construction. The utility will pay nearly all the cleanup cost, but local governments will need to pay for off-site improvements to accommodate the facility unless they successfully bargain with the utility. These are the kinds of impacts reporters would normally cover—that is, the direct environmental and immediate financial impacts.

There are "downstream" and "upstream" impacts. The downstream effects are set in motion by the project. On the side with the proposed facility, the mayor and city council can publicly show that the city's worst brownfield eyesore has been turned into a productive use, and that the city hopes to use that project to attract warehousing and other compatible commercial land uses to adjacent underutilized sites. The city will collect revenues from the utility and expects jobs to follow from the opening of the facility and the other compatible land uses.

On the other side of the river, the most obvious downstream impacts are the noise and air pollution generated by the facility and the visual change it makes to the landscape. Some residents and others who come for fishing or boating may find these effects so objectionable that they choose to leave. Furthermore, a natural gas pipeline is built under the river and onto the less developed side, causing at least temporary congestion and ecological damage on that side of the river.

The upstream effects are steps that are precluded by the decision. In this case, the plant reduces the attractiveness to developers who had hoped to build a marina and ecological preserves on both sides of the river. A ferry service linking the two sides is shelved. Officials on the less developed side are convinced that the stigma of the buried pipeline and visible and unattractive electricity facility preclude their plans for recreational facilities on their side of the river. Plans for condominiums for recreational boaters and bird watchers are cancelled.

The next set of ripples capture some likely additional spatial and temporal effects. Since the utility does not need to fit the electricity peaking facilities into an existing facility at another location in the region, it can design a new baseload unit on that existing baseload site 10 miles away, which means that it need not purchase expensive electricity from other utilities located hundreds of miles away. It also means that the publicity conscious utility agrees to retrofit the existing coal-fired units at this site to reduce emissions as part of the agreement to expand baseline capacity. The net effect will be less air emissions at the site, despite the increase in capacity.

The decision also impacts another utility group located hundreds of miles away, which had hoped to sell its excess baseline capacity from its three nuclear power plants to the local utility. The decision relieves state government and utility officials of the immediate pressure of convincing state residents that they need to purchase more energy efficient air conditioning units that causes the summer peaking problem. The decision shifts the plan for building an ecological preserve, marina, and condominiums 15 miles downstream, where it is welcomed by local government and residents. In other words, if the first facility is built on the brownfield site, residential and commercial facilities can be built at a site 15 miles away, and energy facilities upgraded on another site 10 miles away.

Still another more distant set of ripples is seen 5 years after the new peak facilities and new baseload capacity have been added. The overall state economy has benefited by these additions and in fact this action has been replicated elsewhere in the state, which wants to be more energy independent. The town that added the new peak electricity site has successfully attracted warehousing to the area, thereby boosting the local economy on that side of the river. Personal disposable income has gone up and is reflected in some new housing and retail opportunities for local residents. More residents can afford health care services and gradually children's health and school attendance improves because poverty has decreased.

On the other side of the river, which opposed the facility and lost the preserve, marina, and condominium complex, no new major development has occurred. The community continues to search for a set of land uses that can use its river edge. Elected officials are angry but have no recourse.

The last ripples are harder to measures but are visible in the distance. The energy projects on site and 10 miles away continued reliance on fossil fuels, although the utility did install the best available technologies to limit emissions. If it had purchased baseload capacity from another utility, additions to local electricity grid systems would have been required. This would have meant excavations for some additional above-ground towers through local parkland. Also, with no new local peak load systems constructed, the local area would have been more vulnerable to summer peaks outages, which would have caused new businesses not to build in the area.

In essence, following the ripple effects of decisions requires tracing the likely outcomes of what may appear at first to be straightforward decisions. The reporter who follows the primary story through some ripples in this way will find mostly predictable endpoints, but also some unexpected and counterintuitive outcomes that will be newsworthy. Every brief in this book has ripple effects. However, among the 21, we suggest that "Closing the Civilian Nuclear Fuel Cycle," "Managing the Nuclear Weapons Legacy," "Nuclear Nonproliferation," and "Global Warming and Fuel Sources" have the most local, national, and international impacts and the longest temporal reach.

Worldviews of Technology, Population, Resource Use, and Environmental Degradation

This subsection summarizes a core concern of the authors of this book and of those we interviewed: the relationship between population, resource use, human and environmental degradation, and technology.

When people employ technology to create products, deliver services, and engage in activities, what they do has environmental and human impacts. With the caveat that we are generalizing, degradation processes may be divided into two categories. The first, overusing existing resources, is caused by populations growing too rapidly in areas that lack enough water, soil, fuel, and other resources to provide even modest supplies of products and services. The second category, endangering populations, results from the consumption of enormous amounts of resources per capita. While relatively few people in these nations lack shelter or sufficient food compared with those in less developed countries, the so-called developed nations consume so many resources and emit so many residuals that they potentially have endangered their own populations and arguably everyone else.

For more than 40 years, the United Nations, World Bank, World Resource Institute, Organization for Economic Cooperation and Development, World Watch Institute, and others have warned that we are compromising current populations and future generations. Enacting policies that get as much as possible from each resource, using renewable resources, recycling spent materials, and emitting as little as possible into the environment is essential. The principles of the environmental ethic have been widely touted by some academics, not-for-profits, and individuals who press for pollution prevention and green economic policies. Yet the reality is that management of resources is a complex endeavor involving science, engineering, politics, economics, and social and ethical considerations. In this web of complexity, technology is too often singled out as the culprit or savior for those seeking reasons for failure and success.

The debate about the advantages and disadvantages of new technologies has gone on for centuries, involving vigorous debates and protests about such innovations as skyscrapers; streetcars, trains, and airplanes; and drugs and other medical interventions. Technology debates, we believe, have heated up during the past 20 years because of the exponential increase in scientific exploration and the technologies derived from it.

Not only is scientific uncertainty present in many instances, but research shows that people's ability to perceive the future begins to go dark beyond 15 to 20 years, about a generation. Hence, just below the surface of strongly stated opinions about many technology debates lie two very different worldviews of technology that people revert to when uncertainty about technology and their inability to perceive the future is too great. One involves optimism about the future of technology, the other caution.

Recognizing that the two worldviews presented here are stereotypes that are imperfect fits to real individuals and organizations, we have repeatedly seen information-based arguments fall back to technology-optimism and technology-caution positions. Technology optimists tend to view the world as a place of unlimited resources. Human ingenuity, they feel, will improve public health and the environment, increase wealth and distribute it across the earth, and help solve the world's political problems. They do not believe that we will run out of scarce resources. In fact, they argue that scarcity leads to price increases, which in turn stimulate research, leading to more efficient use of resources, and to resource substitution. For example, as natural gas and oil prices increase, nations will turn more to nuclear power, as well as make more efficient use of fossil fuels.

Technology optimists typically embrace mega-projects that concentrate people and capital to produce massive centralized facilities, such as large clusters of nuclear power plants or gigantic tankers and cargo ships that can transport products across the globe and then offload them in large ports from which they are dispersed on large rail and road networks. Technology optimists do not focus much on population growth because they view people as the key to human, economic, environmental, and political health. They observe that economic development has led to lower birth rates and a stabilized population. In other words, economic development leads to both a decrease in population growth and less poverty.

Their technology-cautious counterparts do not discount human creativity. However, they point to some of our technological creations that have led to human and ecological tragedies. They tend to view resources as finite, they worry about uncontrolled scientific investigations that could lead to problems at a later time, they call for major investments in pollution control and strong regulations, and they worry about population growth. Also, they prefer small-scale technologies to large centralized facilities.

Whereas the optimist group wants to speed up the pace of research, their cautious counterparts prefer a much slower pace of research and demonstration. Some, for example, would argue that increasing reliance on nuclear power means more nuclear waste management in perpetuity and possibly proliferation of weapons grade nuclear materials, arguments that proponents assert are addressed by the projects described in this book. Pointing to Three Mile Island and Chernobyl as illustrations of what is wrong with nuclear power, they argue for a larger number of decentralized facilities that rely on local energy resources, and above all they argue for conservation, which many feel will be undermined by the construction of large centralized nuclear power plants. The issue for reporters and the public is to determine how much the pro and con nuclear arguments are based on science and how much they are grounded in perceptions and values.

In short, large nuclear power plants and waste management facilities are among the most complex and large-scale projects in the world. The opponents and proponents of nuclear power and other nuclear technologies often evoke worldviews of technology as illustrations of their viewpoints. We recommend Simon (1990) as representative of optimism about technology, and Commoner (1971) and Ayres (1998) as representative of concern about the ramifications of technology and unbridled growth.

References

Ayres, R. (1998). *The turning point: The end of the growth paradigm.* London: Earthscan.

Commoner, B. (1971). *Closing the circle.* New York: Knopf.

Sandman, P. (1993). *Responding to community outrage: Strategies for effective risk communication.* Fairfax, VA: AIHA Press.

Simon, J. (1990). *People, resources, environment, and immigration.* New Brunswick, NJ: Transaction.

Covering Nukes: Play Hard, but Play Fair

Written by Tom Henry

The nuclear age is more than 50 years old. Yes. *Fifty.* While the idea of splitting atoms and generating electricity by a chain reaction called nuclear fission is more than a half-century old now, the debate remains hotter than ever.

Are nuclear plants safe, especially as they get older? Can their biggest by-product—spent reactor fuel, the only material in civilian hands classified as high-level radioactive waste—be managed properly? Will the anticipated new breed of reactors be that much better? And, even if they are, will they be too costly to build?

These aren't easy questions to answer, whether you're probing vulnerability to terrorist attacks or mundane wear-and-tear issues, such as the corrosion rate of a certain type of metal alloy, Alloy 600. The latter can be found in most of today's nuclear plants and is not as rust-resistant as once thought.

So how do you really comb through the rhetoric and get past the raw emotion?

Get to know the issues.

School yourself. Learn whom to trust. Stay neutral and hold people accountable.

Remain humble and hungry enough to learn more. Maintain an insatiable curiosity.

Decipher jargon and write with eloquence. Separate science from politics while recognizing that both exist.

Write with flair and passion. Don't get flippant.

Think globally and write locally. Tell people why it matters.

And wish for luck. That's right—luck. Keep looking for that whistleblower or anonymous, inside source who will walk you through the bureaucratic maze you've entered. The stronger your credibility, the better your chances are of landing a key ally. But the reality is you will need to have some doors opened for you. Try as hard as you will, you will still need a certain amount of luck.

But your odds of making that key contact or being directed to that treasure of information in a seemingly arcane and technical document will improve greatly with a good mix of determination and credibility.

A Little Background

There are 104 nuclear power plants collectively generating 20% of America's electricity today. They have one of two types of reactors. About two thirds are pressurized water reactors, known as PWRs for short. The other third are boiling water reactors, known as—you got it—BWRs. The fundamental difference is that PWRs operate at higher pressure and higher temperature than BWRs. PWRs are akin to old pressure cookers found in your mother's kitchen. They're more efficient and powerful, but can be a tougher tiger to tame.

The general concept of nuclear reactors is that they generate steam—i.e., power—to spin turbines that create electricity. PWRs and BWRs go about it slightly differently. With PWRs, the steam is generated by coolant water that has passed through reactors operating as high as 605 degrees. Both types of reactors harness the intense heat created inside the reactor during the fission process, in the form of steam that can spin turbines to generate electricity.

The idea of doing that for peaceful, civilian uses dates to a famous speech former President Dwight D. Eisenhower delivered to the United Nations General Assembly in New York City on December 8, 1953. Called "Atoms for Peace," it is widely regarded as the dawn of the nuclear age.

Today's power stations, of course, didn't get built right away. They took years of planning, arriving after an era of experimental test reactors. And, when you get down to the engineering specifics—and go beyond the two basic types of reactors—there's actually a hodgepodge of 104 different designs across the landscape. Uniqueness makes baseball stadiums interesting. But it also makes nuclear plants confusing not only to the layman but to skilled engineers, too.

The Nuclear Regulatory Commission—the government agency that oversees the nuclear industry—is now in the process of determining which of America's 104 existing plants are in good enough shape to have their licenses extended by 20 years. Possibly even another 40. Such evaluations will continue for years. That same federal agency has been promoting streamlined "cookie-cutter" designs to make parts—and training—at future nuclear plants more interchangeable.

In the fall of 2007, the U.S. Nuclear Regulatory Commission got its first application for a new nuclear plant since the 1970s—a proposed twin-reactor expansion of the South Texas nuclear complex. It's a project that supporters believe could help usher in a new breed of advanced and sophisticated reactor.

America needs more power. The U.S. Department of Energy projects a 50% increase in electricity demand between 2007 and 2037. If true, the nuclear industry would need to build 50 more nuclear plants with today's power output during that time just to continue serving 20% of the market.

Between tracking the relicensing efforts of today's fleet and the next chapter of the industry's evolution, should new plants be built, nuclear power will be in the news for decades to come.

Getting Started

First, don't freak out about nuclear power. Yes, it's incredibly complex. Remember the phrase, *It's not rocket science?* Well, nuclear power is, sort of. But not really. Take it a step at a time and you'll be amazed what sinks in. A few tips:

- The Nuclear Regulatory Commission (NRC) (*www.nrc.gov*) is the federal agency that oversees the nuclear industry. Formerly known as the Atomic Energy Commission, its name was changed in the 1970s when Congress did away with its former advocacy role in favor of strictly regulating the industry. Study this agency's Web site. It has fact sheets on reactor operation to waste disposal. You can retrieve transcripts of speeches and find out about anything from upcoming enforcement hearings to public meetings about nuclear topics at large. Get to know its public affairs officers in the agency's headquarters and in its four regional offices. Familiarize yourself with the NRC's Agencywide Documents Access and Management System (ADAMS), the agency's primary database for public documents. NRC officials admit it can be a little clunky, although it's gotten better. If possible, get ascension numbers for your ADAMS searches, although you can always try to fish around with names. Public affairs officers can help you through it. For more detailed information about how nuclear power fits into the nation's energy picture, contact the U.S. Department of Energy (*www.energy.gov*) and its affiliated Energy Information Administration (*www.eia.doe.gov*). The latter especially is an important resource for journalists; it offers quick statistics on all types of energy sectors and users.
- The Nuclear Energy Institute (*www.nei.org*) is the industry's Washington-based lobbyist. Its site has a wealth of information from the pro-industry

point of view, from speeches to fact sheets to press releases. The organization has public relations officers available 24 hours a day and usually provides quick responses. It is considered the industry spokesman on Capitol Hill. Another good source for getting the industry viewpoint is the American Nuclear Society (*www.ans.org*).

- The World Association of Nuclear Operators (*www.wano.org.uk*) and the International Atomic Energy Agency (*www.iaea.org*) offer a global perspective.
- The Union of Concerned Scientists (*www.ucsusa.org*) is a watchdog group based in Cambridge, Massachusetts, and has members of its Washington office tracking nuclear issues on Capitol Hill. Chief among them is David Lochbaum, a nuclear safety engineer who is widely quoted in the press because of his training in the industry. It is one of the leading groups in the watchdog community. Others include the Nuclear Information and Resource Service (*www.nirs.org*), Greenpeace (*www.greenpeace.org/usa*), and Beyond Nuclear (*www.beyondnuclear.org*), as well as many others.
- An interesting Web site for a panoramic view of the industry, both inside and outside of the United States, is Joseph Gonyeau's Virtual Nuclear Tourist (*www.nucleartourist.com*). It offers backgrounders for the layman, including information designed to help improve the quality of journalism.
- Seek out information from the University of Chicago, the Massachusetts Institute of Technology, the University of Michigan, Ohio State University, and other major universities that currently have or formerly operated research reactors, as well as schools with nuclear engineering specialties.
- Find out what staffers in U.S. House and Senate offices know about nuclear power, as well as the views of the congressmen and senators who employ them.
- Take the time to read a few books. There are too many to list here and, as you might expect, they run the gamut in terms of point of view. One that should be on your desk at all times, though, is the NRC's *Information Digest.* It's a reference booklet that, pound-for-pound, it is one of the more concise and handy things our government publishes. The NRC typically provides one free upon request and sells additional copies. It offers statistics in a pinch, plus it's been expanded to include more visuals, other graphics, and background than its predecessors. You can flip to the back and find out, for example, the date your local nuclear plant went online, the date its 40-year operating license is due to expire, and background on plants that were shut down or cancelled. The NEI also has published a handbook for journalists interested in covering nuclear power, now in its fourth edition.

- Visit a plant. Tours aren't as common as they were before the terrorist attacks of September 11, 2001. But some utilities are doing them. Keep in mind that nuclear plants are owned and operated by private companies, not the government. So be careful about how hard you push, because the call on whether to allow visitors rests with the utilities—not the government. Remember, too, that visiting one plant isn't the same as visiting all 104, because of the varying designs. Utilities typically will have different levels of tours. Ask for the one that gives you the greatest access, the one in which you will suit up and enter the plant's radioactive containment area. The only time you—or anyone else—will be able to enter it is when the plant is idle for refueling or maintenance. And don't be afraid: The containment area should be clean enough that, by heeding procedures, which includes not grabbing or rubbing up against everything, if anything, you are exposed to a level of radiation so insignificant it can barely be recorded. Refuelings are normally the longest outages and the easiest ones for utilities to accommodate tours. They typically occur every 18 months to 2 years, depending on the grade of uranium used in the fuel. So be flexible and plan ahead. Hardly any outsider has ever been known to be given access into the heart of a nuclear plant on a moment's notice. Ideally, the utility would allow you to sit in on one or two of its outage briefings. Although plants are idle during outages, those are actually the busiest time at nuclear plants. Workers are assigned to complete thousands of tasks within a limited amount of time, typically 4 to 6 weeks—the kind of work that can't be done while the plant is operating.
- Go beyond the public affairs officers. Try to find engineers and other outside experts to explain the inner workings of a nuclear plant to you. Take a sincere interest in getting at least a layman's understanding of the technology. Put yourself in their mindset. The NRC has an average of two resident inspectors assigned to every nuclear plant—inspectors who work within the confines of the plant and walk its hallways daily. If possible, get to know them. They can be a tough nut to crack, because the agency typically rotates them to different plants every 2 to 3 years to keep them from getting too cozy with the utility they're regulating. But the resident inspectors are the agency's eyes and ears. They can give you some of the most current and in-depth information. Ask current and former resident inspectors to explain to you how they do their jobs—the kind of inspections they perform. Their findings are typically reported to the agency's regional office, where it may be used by regional administrators or supervisors to brief regional staffers.

- Go to conferences, public meetings, and classrooms whenever practical.
- Go online and find the syllabus of a respected nuclear law, history, or engineering professor.
- Check to see what information you can get about nuclear plants in your state from your state utility radiological safety board, which can often be found through your state health department. It is not a licensing board but provides policy information to your governor. It may go by different names. If it's not affiliated with your state health department, try your state highway patrol or state police.
- Educate yourself about what happened at Three Mile Island Unit 2 near Harrisburg, Pennsylvania, in 1979. Learn what changes occurred throughout the industry. One was the development of emergency planning zones and evacuation plans. Another was the creation of the Institute of Nuclear Power Operations or INPO, an industry-funded, Atlanta-based group that strives to raise the performance level of nuclear plants industry-wide. INPO works only for its member utilities and, as a rule, does not speak to journalists. But it is important to know what it means to the industry. On the watchdog side, you may want to familiarize yourself with Three Mile Island Alert (*www.tmia.com*), a nonprofit citizens group that was created in 1977, 2 years before the half-core meltdown of Three Mile Island Unit 2.
- Learn about organizations and companies that are key players. INPO is one. The Electric Power Research Institute or EPRI, of Palo Alto, California, is another. Framatome, a company based in France, is the world's largest parts supplier for the nuclear industry. Bechtel is a major contractor. General Electric and Westinghouse are major designers. Babcock & Wilcox designed seven reactors in operation today that the industry and the NRC considers to have unique challenges.
- Study up on Nevada's Yucca Mountain, the only spot in the United States being pursued as a possible dump for spent nuclear reactor fuel, the only material in civilian hands classified as high-level radioactive waste. Learn why such a site is important to the future of the nuclear industry: High-level radioactive waste is its greatest obstacle.
- Find out about the transportation routes under consideration for shipping the high-level waste out to Yucca Mountain someday by both rail and highway. Get a copy of the Department of Energy's latest proposal, along with an explanation about how it was developed.
- And don't forget about low-level radioactive waste, which can include just about everything at a nuclear plant other than what's pulled from the reactor core, as well as waste from hospitals, universities, and dental offices.

Find out whether your state is in a decentralized regional compact or where exactly low-level radioactive waste generated within your state boundaries ends up being shipped. There is huge money and not-in-my-backyard political pressures exerted in that issue, too.

- Familiarize yourself with some of the basics of nuclear history—Enrico Fermi, the Manhattan Project, and the 1986 explosion at the Chernobyl nuclear plant near Kiev, Russia, for example. You need this for a little context. The half-core meltdown of Three Mile Island was not only the most tense nuclear event on U.S. soil because of what happened but also for what *could* have happened if people had panicked—and, interestingly enough, because of the communication technology at the time. Believe it or not, there was only one telephone line into the Three Mile Island Unit 2 control room when the event occurred. Even then-President Jimmy Carter had trouble getting through. And don't dismiss the Chernobyl disaster just because it occurred on the other side of the world. I once did a story about how scientists had determined that radioactive fallout from that 1986 explosion was conceivably strong enough to cause or exacerbate a few cases of cancer in the Pacific Northwest. Learn about the near-rupture of Davis-Besse's old reactor head in northern Ohio in 2002, now widely viewed as the second-worst safety lapse in U.S. nuclear history behind Three Mile Island (not just by the activist community but also by the NRC and the U.S. Department of Justice—that event resulted in a record $33.5 million in fines, far more than what was assessed at Three Mile Island). In 2007, the Union of Concerned Scientists published a 40-page report, *Walking a Nuclear Tightrope: Unlearned Lessons of Year-Plus Reactor Outages* (*www.ucsusa.org/clean_energy/nuclear_safety/unlearned-lessons-from.html*), that details a number of significant issues that have affected the U.S. nuclear industry other than Three Mile Island or Davis-Besse.
- Learn about the role of whistleblowers. A *Time* magazine cover story from March 4, 1996, exposing problems with the management–work force relationship at the Millstone nuclear complex in Connecticut, led to national reform of whistleblower laws in hopes of smoothing out the process for workers to report safety concerns. Similar whistleblower issues arose during the 2-year outage of the Davis-Besse nuclear plant from 2002 to 2004, while investigators were at the height of their investigation into how the plant's old reactor head had become so thinned by acid that it nearly blew.
- Familiarize yourself with the terms *safety culture* and *safety-conscious work environment*. *Safety culture* is the most important. It is a largely undefined assessment of the attitudes toward safety within an industrial site—its work

force morale, if you will. Assessing the workplace atmosphere has been done to some degree in the nuclear industry for years, though possibly by other names. The term *safety culture* grew out of the assessments of the Chernobyl workplace following the 1986 explosion there. It has subsequently been applied to other sites, including Millstone in the mid-1990s; Davis-Besse after the near-rupture of its reactor head in 2002; NASA following the explosion of the space shuttle Columbia in 2003, and throughout British Petroleum's five U.S. refineries after a 2005 explosion at the company's refinery in Texas City, Texas, which resulted in 15 deaths and dozens of injuries. It is not a routine assessment.

- Acquaint yourself with some of the more important national laboratories, including the Argonne National Laboratory (*www.anl.gov*) 25 miles southwest of Chicago, which had direct ties to the Manhattan Project. It was the nation's first national laboratory, chartered in 1946, and today remains as one of the U.S. Department of Energy's largest research centers.
- Learn a little bit of history about nuclear financing. Follow the money trail. The lack of new construction over the past 3 decades was not, as the general public believes, directly the result of Three Mile Island. Applications for new nuclear plants ceased coming in months before that accidents simply because of investor uneasiness over cost overruns. The NRC and the Department of Energy have stated that repeatedly.
- Understand why utilities want to keep nuclear plants operating—why it is in their best interest to maintain a safe, efficient plant. The reason? Money. Nuclear power generates in the ballpark of $1 million a day, easily. One engineer walked me through an exercise in nuclear economics, based on 2007 wholesale and retail costs for electricity. He made the argument that the value of wholesale power generated by some nuclear plants in 2007's energy market actually could be $1.5 million a day and that the value of power generated for the retail market actually could be as high as $2.5 million a day. While the estimates may vary, the important thing to remember about nuclear plants is that the old phrase "Time is money" is extremely pertinent to them. It's why utilities fight hard to shave time off refueling outages—events that happen every 18 to 24 months, depending on the grade of uranium in the fuel rods. It once was unthinkable that reactors could be refueled in less than 30 days. Utilities now strive for that as a goal. The "Time is money" concept also is why the busiest time at a nuclear plant isn't when it is operating at capacity. It's when it's idle. That's because utilities want as much, if not all, of the maintenance and inspections that can't be done while the plant is operating to be performed while the new fuel is being loaded.

- Read a few GAO reports about nuclear power, nuclear safety, nuclear security, nuclear waste, and other issues. The GAO changed its name a few years ago from the General Accounting Office to the Government Accountability Office. But it is still the research arm of Congress, and its reports can be found at *www.gao.gov.*
- Become familiar with the workings of and documents produced by the NRC's Office of Inspector General. Accessible through the NRC's Web site, it is the agency's internal watchdog. Congress created it in the 1980s to probe allegations of coziness between the regulator and the industry.
- Learn who some of the key players are in Congress. In the House, U.S. Rep. Ed Markey (D., Mass.) makes no secret of the fact he is a nuclear industry watchdog. U.S. Sen. Harry Reid (D., Nev.) is sensitive about the Yucca Mountain issue because of the potential ramifications for his state. U.S. Sen. George Voinovich (R., Ohio) has emerged as a key ally for the industry, while calling for and participating in hearings about nuclear safety.
- Find out where some of your U.S. representatives and senators stand on nuclear power. Learn what motivates or doesn't motivate them about the industry. Check their campaign finance reports to see whether utilities that own and operate nuclear plants are major contributors to their election campaigns.
- Find out about some of the key legislation for the industry. One piece is the Price-Anderson Act, seen by critics as a government bailout because it caps liability for utilities in the event of a disaster. The nuclear industry claims its needs that protection.
- Learn about some of the materials that have had problems. One example is a product called Thermo-Lag 330, which was supposed to be a fire retardant. Lab tests showed it burned far more quickly than it was supposed to burn, causing plants nationwide to either remove or install a double-barrier of the stuff.
- If there's a nuclear plant in your area, learn the history of it—how the land in which it was built on was chosen, tested, and approved; who funded and who designed the facility; where it draws and discharges water; how local wildlife have fared; the amount of property taxes the plant has generated for state government or the local school district; how many people the plant has employed at various junctures over the years, as well as the amount of work outsourced to contractors compared to years ago; the strength of the utility's stock, the plant's performance record, and so forth.
- Know what a Condition Report is. An Unusual Event. A Generic Letter. A Demand for Information. And some of the other NRC-industry jargon. Weave through the bureaucratic maze. A condition report, for example, is

document at the worker level—a tip, if you will—that something is amiss and should be checked out. An unusual event is just what it sounds like: something out of the ordinary. But it's an event reported by the utility, not so much an ongoing condition. A generic letter is something the NRC issues on occasion—not with any great frequency, though—when it has reason if there is the potential for an industry-wide problem to address. The agency typically is trying to get its arms around a design or operational issue to see how prevalent it might be. A demand for information is a more serious gesture by the NRC and more likely to be specific to one plant. The agency wants to zero in on a potential problem and is letting the utility know that failure to cooperate could result in serious consequences, up to and including license revocation.

Hold 'Em Accountable

In short, don't get psyched out by the pro-nuke/anti-nuke rhetoric. Yes, it's thick, annoying, and enormous. But the best sources from either side of the nuclear fence will recognize you for doing your job as a journalist: holding people accountable. You'd be surprised, for example, just how much some people in the nuclear industry want their colleagues to be held accountable in the media when they've done something wrong. That's because they don't want their industry needlessly getting black eyes.

An NEI spokeswoman once told me the industry likes me because I'm "tough, yet fair." "We know you're going to ask the hard questions that need to be asked, but you'll be fair about it and will do your homework," Thelma Wiggins told me.

I've received similar sentiments from the activist community which, hopefully, means I'm performing with some sort of fairness and balance while remaining bold and gutsy. As Jim Wojcik, the adviser for my old college newspaper would say, "There's no substitute for credibility."

So go at it. Do your nuclear research. Jump head-first into the topic, seek advice, educate yourself, and enlighten your audience about this form of energy at a critical time in the nation's history.

Play hard. But play fair.

Part II: Briefs

Section 1

Radionuclides and Human Health Effects

Written by Michael R. Greenberg and Bernadette M. West, based in part on interviews with Kathryn Higley, Michael Stabin, and Michael Gochfeld, with comments by Niel Wald

Background

From a young age many people have been taught that atoms form the basic building blocks of matter. In our everyday world atoms are the indivisible piece that defines elements—such as sodium, chlorine, oxygen, and hydrogen. Combinations of these elements go on to build chemical compounds—such as water (H_20) or table salt (NaCl).

When we deal with issues of atomic and nuclear physics, we need to understand that atoms are actually built from multiple components: neutrons, electrons, and protons (and even smaller particles—but that's for another discussion). Electrons orbit at a considerable distance around a positively charged nucleus that contains approximately equal numbers of protons and neutrons. The components of the nucleus are in a delicate balance. The protons, which are positively charged, repel each other (opposites attract and likes repel). The neutrons, which have no charge, act as a buffer—they are the shield between the positive charges and so allow multiple protons to be packed into the nucleus. However, if the proportion of neutrons to protons is not exactly right, or if a nucleus gets bombarded by other particles, this balance is disturbed. The nucleus then seeks to move to a more energetically favorable (stable) state. A chemical element in such an unstable form is called a radionuclide. It is also referred to as a radioisotope.

As these radionuclides rearrange their internal structure, they often give off radiation—energy traveling through space. The radiation emitted from a nuclear rearrangement can be in the form of either waves (like light) or particles (like electrons or helium atoms). When unstable atoms give off energy in the form of particles or waves, the process is called "radioactive decay" and what is emitted or given off from this process is called "ionizing radiation." When ionizing radiation strikes anything—wood, iron, the human body—it creates electrically charged particles called ions, which can have effects on matter, including living things.

Some radionuclides take a very long time to undergo this nuclear rearrangement. For others, the process is very quick (measured in nanoseconds). We use the term *half-life* to measure the time required for half of the atoms of a particular type to lower their energy level through nuclear rearrangement. In other words, a half-life is the time it takes for the material to lose one half of its radioactivity. The radioactive decay process continues until a nonradioactive stable element forms.

To understand radioactive decay, it is important to keep in mind that radionuclides occur naturally but they can also be produced artificially (by people). Naturally occurring radionuclides can be sorted into three categories: primordial radionuclides, chain decay radionuclides, and cosmogenic radionuclides. Primordial radionuclides have extraordinarily long half-lives. Among these are uranium-238, thorium-232, and potassium-40. They have been on the earth since its formation from interstellar dust and were produced from supernova and reactions taking place in the interior of stars. Uranium is the oldest natural radioactive element. It has a half life of 4.5 billion years. This means that a given quantity of uranium-238 will lose half of its strength in 4.5 billion years. It will lose half of its remaining strength in another 4.5 billion years. Chain decay radionuclides are the radioactive elements that result from the decay of several of the primordial elements. They often have very short half-lives. For example, radon-222, a radioactive gas that is a product of uranium-238, has a short half-life of 3.8 days. A classic example of radioactive decay is uranium in potassium uranyl sulfate, which emits alpha particles. This decay results in thorium, which itself is radioactive and emits beta particles. The third group, cosmogenic radionuclides, are continually being produced because of the interaction with cosmic radiation in our atmosphere and surface soils. For example, carbon-14 is continuously formed in the atmosphere by cosmic rays.

Radionuclides can be produced artificially in a variety of ways—in nuclear power plants during the process of fission (when the nucleus of an atom is split), in particle accelerators, and through the detonation of nuclear weapons. Accelerators speed up particles to very high energies and then smash them into specially constructed targets, producing "bits" that provide insight into the very composition of matter. The world's largest particle accelerator, CERN, lies on the outskirts of Geneva on the French and Swiss border. Originally, accelerators were used only in high energy physics laboratories like CERN (or in college physics classes). But now they are often constructed for less exotic applications—to manufacture radionuclides for use in medical procedures.

Most hospitals in mid-size cities will have access to radionuclides produced in accelerators.

The radiation that is emitted in nuclear decay involves electromagnetic waves and particles: gamma rays (γ), which are neutral but highly energetic; x-rays (like gammas, but slightly lower energy); alpha particles (α), which are positively charged; beta particles (β), which are negatively or positively charged; and, neutrons, which have no charge but have almost the same mass as a proton. These types of radiation are the ones people typically encounter. In addition, there are some other more exotic types of radiation that can occur, but typically only when accelerators are operating. Some of the characteristics of the four major types of radiation and the radionuclides that provide a source of that radiation are presented in Table 1.

Among the several different types of instruments that can measure emissions and identify the radionuclide that emitted them is the Geiger counter or Geiger-Mueller counter. (Hans Geiger was an early 20th-century German physicist.) In addition to these instruments, nuclear physicists and engineers have built cloud chambers that allow them to view radioactivity in the same way we see jet contrails in the sky.

In measuring radiation, different units are used depending upon whether you are talking about the rate at which material decays, the risk of exposure to a person, or the radiation dose absorbed by a person. The rate at which radiation is being given off by a radioactive material is measured in *curie* (Ci) or the international unit *becquerel* (Bq)—named after Marie and Pierre Curie, who found additional radioactive elements in uranium ores, and Antoine Henri Becquerel, their colleague, a French chemist who discovered that some elements, especially some heavy elements (e.g., uranium), spontaneously emit radiation. The biological damage from exposure to a person is measured in *rem* (roentgen equivalent in man, which is the exposure in roentgens multiplied by a constant that takes into account the type of radiation). The rem equivalent in the international unit is *sievert* (Sv). The radiation dose absorbed by a person is measured in *rad* (radiation absorbed dose) or the international measure of *gray* (Gy). Prefixes are used to describe smaller fractions of these quantities. For example, 1 rad = 1,000 millirads (or mrads).

The following equivalents help put the different units in perspective:

- 1 Ci = 3.7×10^{10} disintegrations per second
- 1 Bq = 1 disintegration per second
- 1 rad = 100 ergs per gram of tissue
- 1 Gy = 1 joule per kilogram of tissue = 100 rads

Table 1. Major Types of Radiation

Waves

Gamma (γ)

Charge: Neutral

Source: Nucleus of unstable atoms; also accompany emission of alpha and beta radiation during decay

Methods of detection: Survey meters that probe with sodium iodide

Damage/effects: Capable of penetrating most materials, including human tissue; able to cause serious damage when absorbed by living cells, including DNA alteration

Energy level/penetrating ability: High energy compared to visible or radio waves; able to travel many feet in the air and many inches in human tissue; highly penetrating

Protection: Clothing will not protect; shielding requires dense materials (lead, concrete, or several feet of water)—the higher the energy, the thicker the shielding required. Sealed machines that emit gamma radiation are mainly external hazards.

Examples: Iodine-131, cesium-137, cobalt-60, radium-226

X-rays

Charge: Neutral

Source: Electronic part of the atom; x-rays also accompany emission of alpha and beta radiation during decay

Methods of detection: Geiger-Mueller (GM) probes or scintillators

Damage/effects: Less penetrating than gamma rays

Energy level/penetrating ability: High energy compared to visible or radio waves; able to travel many feet in the air or inches in human tissue; highly penetrating

Protection: Clothing will not protect; shielding requires dense materials (lead, concrete, or several feet of water). Sealed machines that emit x-rays are mainly external hazards.

Examples: N/A

Particles

Beta (β)

Charge: Negatively or positively charged

Source: Ejected electrons from orbits or connected internally and orbits electrons

Methods of detection: GM probe (some low energy particles harder to detect)

Damage/effects: Able to penetrate skin to where new skin cells are produced; prolonged exposure on skin surface can cause damage; harmful if ingested

Energy level/penetrating ability: Light; able to travel short range; moderately penetrating ability

Protection: Clothing provides some protection; can be shielded by thin layer of aluminum or several centimeters of air

Examples: Strontium-90, carbon-14, tritium H-3, sulfur-35

Alpha (α)

Charge: Positively charged.

Source: Ejected helium nucleus

Methods of detection: GM probe

Damage/effects: Most are not able to penetrate skin; harmful if inhaled, swallowed, or absorbed through an open wound. Alpha radiation, which is heavier and carries more electric charge, causes more damage than either beta or gamma radiation.

Energy level/penetrating ability: Heavy; low energy; able to travel very short range; low penetrating ability

Protection: Can be shielded by paper or thin layer of water or dust; cannot penetrate clothing

Examples: Radium, radon, uranium, polonium, thorium

To account for differences between various types of radiation (alpha is heavier, carries more charge and causes more damage than beta or gamma radiation) and to be able to give dose from all types of radiation a common measure, the term dose equivalent is used and reported as rem. Scientists calculate dose equivalent by multiplying the absorbed dose (in rad) by a "quality factor" for the specific type of radiation. This is then reported in rem. For example, if a person receives a dose of gamma radiation, the damage experienced will be less than for someone who receives the same dose from alpha particles. Alpha particles will cause three times more damage than gamma rays and therefore have a "quality factor" of 3. The relative biological effectiveness (RBE) of alpha particles varies, depending on the type of effect in question. To protect against the carcinogenic effects of chronic, low-level radiation, the International Commission on Radiological Protection has recommended a weighting factor of 20 for alpha particles.

In the 1970s, as the result of an international agreement, most of the world adopted a new system of measurement. The International System (SI), which uses gray (Gy) and sievert (Sv) for absorbed dose and dose equivalent, respectively, is now the official system used internationally for radiation measurement. Experts in the United States, however, use both the conventional and the new, SI system. The following table shows the equivalent measures in the two systems.

1 Sv = 100 rem	1 rem = .01 Sv
1 mSv = 100 mR (mrem)	1 mR = .01 mSv
1 Gy = 100 rad	1 rad = .01 Gy
1 mGy = 100 mrad	1 mrad = .01 mGy

Dose rate means the amount of radiation a person is exposed to or the amount of radiation absorbed by the body per unit of time. Dose rate is similar to prescription medicine. If you have a prescription that says take 1 tablet daily and you decide to take 3 tablets every third day, you are increasing your dose rate, even though the total dose in 30 days would be the same. If you took all 30 tablets on 1 day, you would have a very high-dose rate.

Identifying the Issues

Like many things in our world, radionuclides—and the radiation they emit—can under some circumstances be hazardous but the level of risk they present varies. Hazard is the potential to cause harm; risk is the likelihood of harm.

Risk varies with the circumstances. A highly hazardous substance that is sealed may present little risk to a person, whereas a rather innocuous substance, such as flour in the form of airborne flour dust in a bakery, may pose a serious risk for a baker with occupational asthma. An agent of relatively low hazard can present substantial risk; conversely, an agent with a high hazard might present no measurable risk in certain circumstances

The degree of risk involved with radionuclides depends on three conditions of exposure: how much, how long, and what part of the body is involved. In trying to understand the risks of exposure to radiation, it is important first to keep in mind that everyone is exposed to some level of radiation. Radiation is part of the natural environment. We come in contact with radiation from materials in the earth's crust, from naturally occurring radiation in the air, and from cosmic radiation from outer space. We have small amounts of radioactive materials in our bodies simply because our environment is naturally radioactive, and we receive radiation exposure as a result of activities we engage in, such as taking an airplane flight. Natural background radiation is part of our lives.

Radiation from natural and man-made sources can be found everywhere in the world. "Background" radiation comes from space (i.e., cosmic rays) and from naturally occurring radioactive materials in the earth and in living things. How much exposure to radiation people receive depends on where they live, their lifestyle, the foods they eat, and their type of housing. In addition, medical tests may also contribute a certain amount of exposure to radiation.

People living at high altitudes have higher exposure rates than people living along the coast at sea level because there is less air above them to block radiation. In an airplane, higher elevations increase exposure levels. Over two thirds of exposure to natural sources of radiation comes from radon—a gas that occurs when uranium decays in the soil. Radon is found across the United States. People are exposed to different amounts depending on where they live and the geological formations in that part of the earth—the closer the bedrock is to the surface, the higher the concentration of uranium. Radon in the ground usually moves up into houses and other buildings through cracks in the foundation. If you live in a tightly sealed building (windows are kept closed allowing no ventilation), your exposure level can be higher. Other natural sources of exposure can come from foods (some fruits and vegetables contain potassium-40, a naturally occurring isotope) and drinking water that contains radon.

Man-made sources of radiation exposure include television screens, computer monitors, smoke from cigarettes, and certain medical tests, such as x-rays and CT-scans. Personal lifestyle factors such as passing through luggage screening at the airport, having smoke detectors in the home, and using gas camping lanterns can contribute to exposure levels. Living in close proximity (within 50 miles) to either a nuclear power or a coal-fired power plant can increase exposure to radiation. Nuclear weapons test fallout is another source of radiation exposure. Exposure from this source has decreased since 1962 when atmospheric testing stopped in the United States.

Approximately 80% of the exposure people receive comes from natural sources in the environment, including radon (~55%), cosmic radiation (6%), radiation from rocks and soil (8%), and radiation found inside our bodies (11%). Man-made sources such as x-rays and nuclear medicine contribute 11% and 4%, respectively. Commercial applications of ionizing radiation contribute another 3%, including sources such as the domestic water supply, building materials, mining and agricultural products, and fuels—in particular coal—and other, smaller man-made sources, such as television sets. Production of electricity by nuclear power adds about .001% of our exposure and electricity from coal-fired power plants another .005% on average.

The Health Physics Society estimates that external background radiation exposes the average person in the United States to approximately 60 mrem per year. Radioactivity from within the body exposes the average person to 40 mrem per year. While there is variation depending on where a person lives, radon in the home exposes the average person to 200 mrem per year and medical x-rays expose people on the average to 60 mrem per year. For people who work with or around radioactive materials, international standards place exposure limits higher—at 5 rem or 5,000 mrem (50 mSv) because the work provides a compensating benefit to the worker, that is, a paycheck.

How do radionuclides cause harm? When a particle undergoes radioactive decay, it can emit a variety of charged particles and electromagnetic radiation in the form of waves (like visible light, only with much higher energies). If this decay takes place inside living tissue, these emissions have the potential to cause chemical reactions that can damage the cells. Our cells have very sophisticated repair mechanisms that have been designed to cope with the type of damage caused by nuclear radiation. If too much damage is caused, however, these repair mechanisms may be overwhelmed. If the cells cannot be repaired, they may die; too many dead cells can lead to organ failure (and death of the individual). Conversely, if the cells do not die but are "misrepaired," the damaged cells may go on to become cancerous.

People can be exposed to radiation by swallowing or breathing in radioactive materials or particles or by having direct contact with them. For example, people can swallow radioactive particles in water that contains alpha emitters from certain minerals. These particles can become lodged inside the body, exposing the organs and bones to radiation. Radioactive materials and particles people breathe in can lodge in the lungs, where they decay and cause exposure. People can inhale radiation, for example, from contaminated soil that has been disturbed. Some industrial processes, such as incineration, can release radioactive particles into the air or create ash that can be inhaled. Radon in the air is considered more dangerous than radon in drinking water.

When a person comes into direct contact with a radioactive material, radioisotopes in the form of gas, liquid, or solid can settle like dust, covering the skin, hair, eyes, and clothes. Removing the clothing and washing off the contamination can stop the exposure. Alpha particles can enter through a cut, while beta particles can burn the skin and damage the eyes. Exposure to gamma radiation is more serious. In a case of direct exposure, it is important to determine how long the person was exposed, how close he or she was to the source, and whether anything was between the person and the source that was capable of absorbing the radiation.

How harmful is radiation? As noted earlier, we have grown up being bombarded with radiation from the sun and from the decay of radioactive substances in the air, soil, and water around us. Some scientists even assert that a very small amount of radiation exposure improves health by stimulating the immune system—this is called hormesis. This view, however, is controversial and not accepted by the majority of scientists working with radiation, who attribute the primary basis for an adaptive response to the capacity for repair of DNA damage. At one end of the scientific spectrum are scientists who believe that a single radioactive particle can cause cancer. At the other end, a common view held by specialists in the field is that a healthy person's immunological system can cope with the low level of damage caused in cells by radioactive decay that occurs from environmental sources, or even somewhat higher amounts—that is, there is a threshold before irreparable damage occurs.

Damage to the DNA of a cell may result from a single ionizing ray or particle crossing its nucleus. Such damage may not be accurately repaired at high doses. The probability that the residual damage may ultimately be expressed in the form of a given mutation or chromosome aberration is small. The long-standing disagreement is at low-dose exposures. Can a single ionizing ray or particle cause cancer? Renowned scientists disagree, including reviewers of this volume. There is on-going scientific debate about whether the relationship observed in human populations exposed at high levels can be interpolated linearly downward to zero. At low-dose exposure levels, there are natural repair mechanisms in the body but the details of these mechanisms and their effectiveness to completely mitigate damage is the subject of current research and debate. The linear no threshold hypothesis (LNTH) assumes that the risks for cancer at high doses extrapolate linearly to low doses. According to the LNTH, the chance of getting cancer is directly proportional to the radiation dose, continued all the way down the line, and therefore even extremely low levels of exposure could cause a cancer. Only zero dose equals zero risk. At extremely low-dose levels, the data are weak and it is difficult to see what is happening. For regulatory purposes, this seemingly conservative view has been used.

The data do, however, provide the basis for the view that the risks of genetic and carcinogenic effects of radiation are probabilistic, or stochastic, statistically related to the amount of the dose, and can therefore be expected to increase in frequency as nonthreshold functions of the dose. This risk of genetic and carcinogenic effect is in contrast to acute radiation sickness and other usually acute tissue reactions, which are known to be produced only by doses that are large enough to kill substantial numbers of cells in the affected organs. Viewed in this context, the risks of carcinogenic and genetic effects are expected to be increased by a given dose, whether it is received in a single, brief exposure or over a prolonged period. The magnitude of the increase depends on the linear energy transfer within the impacted range (LET) of the radiation, the dose rate at which the transfer takes place, and other factors, including the particular tissues and cells affected.

There are two types of radiation exposure effects depending on the duration of the exposure (short- or long-term) and the level of exposure or dose: (1) short-term, high-level exposures and (2) long-term, low-level exposures. Short-term, high-level exposures produce noncancerous health effects that appear quickly. They include burns and radiation sickness, also called "radiation poisoning," which can produce nausea, hair loss, burns to the skin, muscle weakness, and reduced organ function. When medical patients undergo radiation treatment, they experience some of these acute effects because of the relatively high exposure they receive in their treatment.

At high levels of dose, clearly observable phenomena occur, such as skin burns, evidenced by redness, burns, and ulceration. This effect becomes more severe the higher the dose of radiation. Typically there is a threshold below which a person is not going to show any damage at all; above this level, everyone will probably get some damage. Depending on how high above that level one goes, and the size of the area involved, a person will show more and more skin changes, including hair loss and other tissue and organ damage, and in rare cases, death occurs. With acute exposures the impact may not necessarily occur immediately. If a person receives a skin burn from a high exposure to radiation, it may not become obvious for several weeks and hair loss may not begin for 2 to 3 weeks. Most acute exposures involve injuries that heal themselves or benefit from dermatological treatment, although higher exposures may damage the underlying blood vessels, interfering with recovery of the skin. Death is an extremely rare outcome because exposures are almost always low dose.

The lethal dose 50 (LD50) for penetrating radiation is the dose of radiation that will kill half the exposed population. The LD50 is approximately 3.5 Gy for untreated patients and 5 Gy for those receiving full medical treatment.

There are three stages of health effects for acute radiation exposure. The higher the radiation dose the shorter the time of onset of each stage and the more severe the signs and symptoms—allowing for individual variations and the fact that not all patients will experience all symptoms. In the first couple of hours to the first couple of days after exposure (known as prodrome), the person may experience nausea and vomiting, followed by malaise, fatigue, and weakness. A latent period follows the prodrome, during which the symptoms subside and only weakness and fatigue are still present. Eventually with the manifest illness, the clinical signs and symptoms of damage to major organ systems, including the cardiovascular and central nervous systems, become apparent. The molecular cause of the resulting disease is damage to the DNA, and sepsis is usually the cause of death.

Acute radiation syndrome is actually several subsyndromes that evolve and are cumulative with increasing dosage. They are referred to as the hematopoietic subsyndrome, the gastrointestinal (GI) subsyndrome, and the cardiovascular and central nervous system subsyndrome. The hematopoietic system of organs and tissue is involved in producing blood. The hematopoietic subsyndrome results from doses > 0.7 Gy and affects the stem cell of bone marrow cell lines, reducing or stopping blood cell production. This subsyndrome is marked by mild weakness, fatigue, and anorexia at first, followed by possible hair loss depending on levels, weight loss, and bone marrow atrophy, hemorrhage, and infection. Treatment for cases with few complications involves preventing infection and hemorrhage and restoring blood-forming elements. GI subsyndrome occurs at absorbed, whole-body doses > 5 Gy. The latent period that follows prodrome is marked by disabling malaise and weakness, followed by mucosa breakdown, abdominal distention, vomiting, diarrhea, and GI collapse. Cardiovascular system and the central nervous system subsyndromes begin at doses > 20 Gy with full syndrome present at exposure levels > 50 Gy. Prodrome can begin in as little as 5–10 minutes, with rapid onset of uncontrollable nausea, explosive vomiting and diarrhea, epileptic seizures and altered mental status, eventual loss of consciousness, increased intracranial pressure, respiratory distress, and swings in blood pressure. Coma and death occurs within 2–3 days.

Health effects sometimes occur as the result of long-term, low dose exposures. Although cancer is caused by multiple factors, it is a known health effect associated with some levels of radiation exposure, and, to a lesser extent, hereditary effects. Hereditary effects from parental genetic damage in humans are discussed less in the literature because so far they have not been proven to have occurred. With regard to the long-term health effects of chronic exposures, the relationship is much harder to determine because there is a latency period. Effects from the dose may not be observed for many years—sometimes 20 years or more. At that point, if a person does get a cancer from a radiation dose, it is much harder to attribute the cancer to radiation because populations who were exposed 20 years before have since moved around and been exposed to many other hazards. Also, since radiation exposure does not produce unique forms of cancer, only an increased frequency, the temporal relationship of its appearance to the time of exposure and the frequency of occurrence to the magnitude of the exposure in a population are used to determine the likelihood of a relationship of radiation exposure to various forms of cancer. The following cancers are linked with radiation exposure (National Research Council, 1990):

- bile ducts
- bone
- brain
- breast (male and female)
- central nervous system
- esophagus
- gall bladder
- kidney
- leukemia (except chronic lymphocyte)
- liver (except with cirrhosis or hepatitis B)
- lung bronchus
- multiple myeloma
- non-Hodgkin's lymphoma
- ovary
- pancreas
- pharynx
- rectum
- renal and urinary bladder
- salivary glands
- skin
- small intestine and colon
- stomach
- thyroid

The calculation of the likelihood of outcomes from chronic exposures is based on expert assessments of the risks of low-level irradiation in human populations. These assessments are based on the integrated analysis of all relevant epidemiological and experimental data, and not just on extrapolation from human data alone.

What Are the Risks?

Some risks are increasing and some are decreasing. With regard to peaceful uses of nuclear materials—nuclear power plants, dental x-rays, and many other uses of radioactive nuclear materials—people are safer than they were 2 decades ago. For example, in the United States, patients undergoing dental x-rays are protected by a lead vest, the calibration of equipment is more accurate, and technicians leave the room before taking x-rays. New nuclear technology often emits radiation that is less than background levels. And yet, there are more exposures because of greater use of x-rays, CT scans, and nuclear medicines and an increasing frequency of self-prescription. Overall, there are many more small exposures to low levels of radiation than there were a generation ago.

Globally, the risk from nuclear weapons for some people and places seems to have increased, while it has decreased for others. The United States and Russia have reduced the number of nuclear weapons in their arsenals. Yet nuclear weapons proliferation is a major concern; the United States and Russia continue to have more than enough nuclear weapons to kill and injure many millions of people. (See "Nuclear Nonproliferation.") Managing waste from these weapons is a major long-term challenge. (See "Nuclear Waste Policy.")

Scientists, engineers, educators, and policy makers can collectively reduce risk from radiation exposure by various methods, such as shielding. With regard to health science, we have learned a great deal about human reaction to radiation by studying cells, animals, atomic bomb survivors, uranium mine workers, and other heavily exposed groups. We have a much better idea of how to protect people who work with radioactive materials and how to confine radioactive materials. With regard to technology, engineers are pursuing safer and more accurate medical devices, nuclear power reactor designs, and methods of managing and reusing nuclear waste. Some of these efforts are described elsewhere in this volume.

The United States is a major user of nuclear materials. Thousands of federal and private workers are employed at the U.S. Department of Energy's nuclear laboratories, weapons facilities, and locations where nuclear waste is managed, and the U.S. Nuclear Regulatory Commission is charged with protecting the public and workers. Despite this organizational infrastructure, the United States has fallen behind several other countries in the training of its nuclear work force and that work force is aging. Recent reports by the Oak Ridge Institute and the Department of Energy conclude that if the United States is going to increase its reliance on nuclear power, it will need to invest funds in educating the next generation of nuclear scientists, engineers, and technicians.

What Reporters Need to Know

Reporters can better inform their audiences by asking *risk-related* questions before writing their stories. Not all of the following questions are relevant to every exposure context.

1. How dangerous are the radioactive substances? What do they emit? What is their half-life? Are the radioactive elements contaminated with other substances? What is the implication of any contamination?
2. How much of the radioactive material is present? How does this compare with other common sources of radiation and radioactive material?
3. How effectively is it contained? Is it in a solid form that reduces its mobility? Is it enclosed in a tank or encapsulated in a solid form that would limit its mobility? Is it in a liquid form, in tanks below or above ground? Is it in a gaseous form? If so, how is it being contained?
4. If the substance escapes containment, how rapidly will it be diffused? Will it diffuse through the air, water, or the ground?
5. How many people are nearby who can be potentially exposed? What animals, plants, and other parts of ecological systems can be affected by emissions?
6. Can people and animals be relocated so that they are not exposed during an emission? Where will exposed and injured people and animals be taken? How rapidly can a replacement water supply, if required, be secured? How will contaminated structures, property, and soil be managed? Who will be responsible?
7. How will the area be secured until it is safe? How long will the public be kept out of the area? Will access be limited in perpetuity?
8. How will government and responsible parties follow up so that the chances of subsequent exposures are reduced?
9. What can people do to protect themselves from harm?

A nuclear weapon detonation or a major nuclear power plant explosion and fire, such as occurred at Chernobyl, can kill or injure anyone because of thermal, blast, and radiation effects. However, by studying the victims of Hiroshima and Nagasaki and other exposure incidents, we know that some people are more sensitive than others. The young, elderly, and people with genetically impaired immunological systems are at highest risk. In the repair of cellular radiation injury, some people may not have the capacity for repair of damage to DNA. Our experts note that a small proportion of people do not have the normal capacity to repair cells. Accordingly, they may exhibit increased sensitivity to x-rays and be at higher risk. They should minimize exposure to CT (computed axial tomography) scans, diagnostic x-rays, and other radiological treatments and tests because of the doses involved.

Pitfalls

Can exposure to a single radioactive molecule cause cancer? The answer to this question has been debated for decades and has taken on a public and political life of its own, sometimes leading to public confusion. Theoretically, exposure to a single particle can cause cancer, just as exposure to a single whiff of tobacco smoke can cause lung cancer. Realistically, with regard to radiation, we are continually bombarded in our natural environment by radiation and radioactive elements from outer space and other sources on the earth. The body is typically able to repair the damage of a single radioactive molecule—and more. Even people with defective immunological systems are able to repair some damage.

The single-molecule issue too often polarizes political debates about radiation, driving discussion to unrealistic extremes. When federal agencies and private organizations decide how to model the effects of exposure, some models assume that any exposure can cause cancer, that is, there is no threshold. Others assume that there is a threshold exposure that must be reached before irreparable damage occurs.

Determining the scientific basis of any argument is critical in debates involving radiation. If an argument is offered for no threshold, or for some threshold, ask for the scientific basis of the claim. Was it human studies? Animal studies? How many studies and over how long a period? Who paid for them? Were they subject to peer review? Have the results appeared in a peer-reviewed journal? Did the analyst calculate the threshold or lack of threshold from actual data? Or does the threshold or lack of one represent an extrapolation (that is, extending the data beyond the collected data to derive an estimate)? Get answers to these important questions before adopting an explanation.

While some scientists may be reluctant to say there is a threshold, some do not believe that extremely low levels of radiation are harmful and may even be beneficial in stimulating the body's repair system. The question then becomes what's very low, what's moderate, and what's high? That is a matter of discussion and belief, and that is why it is hard for reporters and the public to understand this phenomenon. It is a constant challenge to clarify this issue, make risk estimates reliable, keep them as free of bias as possible and make this issue understandable to the public. In 1990, a group of scientists was convened to assess the biological effects of low-level exposures (BELLE) to chemical agents and radioactivity. The members recognized that while most toxicological studies focus on high level exposures, most human exposures occur at relatively low levels. Consequently, risks at low levels are estimated by various means, frequently making assumptions about which there may be considerable uncertainty. For information about this meeting and newsletters of interest, see the Web page *www.belleonline.com.*

There are few circumstances where the public would be likely to face acute risks (high doses of radiation over a short span of time). The only time one would likely see acute exposures that could produce severe radiation injury or risk of death in the general population would be in a terrorist situation involving a highly radioactive dirty bomb or if a nuclear weapon were detonated. (See "Managing the Nuclear Weapons Legacy" and "Dirty Bombs.") These have been extremely rare and are, one would hope, not likely to happen. So it is critical to establish whether you are focusing on an acute high dose or a dose that is more likely to be repaired by the body. One way is to ask about the dose due to the exposure compared with general exposure in the area. A 10 millirem dose given to a person for a chest x-ray examination for medical purposes should not be compared with that of a 10 millirem dose received by a radiation worker in a nuclear power plant over a 30-day period. The x-ray was given to a small portion of the body and delivered in a fraction of a second. The exposure for the worker was to the entire body distributed irregularly over many days. These are different events and this comparison would be misleading.

In dealing with uncertainty in knowledge and trying to determine the effects of small radiation exposures, the regulatory tendency is to be conservative and overestimate rather than underestimate risk. When the health of the public lies in the balance, and when a regulatory agency is confronted with a range of choices in interpreting the scientific findings, the inclination has been to err on the side of caution and choose the more conservative position. That is, the risk is made to appear greater than it may actually be on the basis of observable effects at high doses. The assumptions then—if based on error—produce dose limits that may be lower than actually necessary. It is difficult to calculate the potential health and economic cost of overly restrictive regulation. If there is a confirmed exposure, it is prudent to probe the likely impacts by asking what the data actually show regarding health effects.

Tying dose to risk can be problematic. Each situation involving a radiation dose produces a unique ratio of risks to benefits. For example, risk benefit ratios vary dramatically for the patient receiving a radiation dose from a medical procedure that is going to benefit him or her directly as opposed to a situation involving a low-level radiation dose from nuclear weapons disposal or nuclear power generation. In the former, the patient willingly receives a certain amount of radiation and has discussed the risks and benefits with his or her health care provider. In situations involving nuclear waste disposal and nuclear power generation, benefits are dispersed far beyond the region where the radiation doses are received, complicating calculation of a risk benefit ratio.

There are dose calculators available on the Web that calculate dose levels. Their usefulness depends on who put them together and how accurate the science is behind them. They calculate dose by automatically attributing a certain amount of mrem for cosmic radiation, average internal radiation from food, water and air, and weapons test fallout, and then they add varying amounts based on where you live, what your house is made of, and whether you wear a plutonium powered pacemaker (replaced by lithium batteries in 1970s), have porcelain dental crowns or false teeth, or wear a luminous watch, how many x-rays or nuclear medical procedures you had last year, how many jet miles you fly each year, whether you have your luggage screened, and whether you watch television, have smoke detectors, use computer monitors, or use gas lantern mantles for camping.

Journalists and the public tend to worry a great deal about radionuclides that are artificially made, such as plutonium and cesium-137. These can be dangerous, but they are intended to be controlled as part of government policy. If they are properly managed, the immediate risk is minimal. Although it is important to ask about how they are managed, reporters sometimes report the *potential hazard* without reporting the *likely risk.* In contrast, naturally occurring radionuclides such as radon and potassium-40 are treated as if they are harmless. But, in fact, we know that some houses were built from materials that have naturally occurring radioactive materials in them and some of them have been demolished or remediated as a result of the risk (e.g., in Climax–Grand Junction, Colorado, and Tiger Bay, Florida). We also know that naturally occurring radon gas can accumulate in unventilated places and, in the worst case, can lead to high levels of exposure and potentially to lung cancer. The U.S. Environmental Protection Agency estimates that up to 20,000 lung cancer cases a year are a result of radon exposure in homes. The important point is that the story should be based on the risk posed within a specific context (see questions under the earlier subhead "What Reporters Need to Know"), and not on the origin of the radionuclide. It is better to avoid the temptation to label radioactivity as good or bad. Instead it is more accurate to clearly state what circumstances make it hazardous or beneficial in a specific context.

Technical mistakes can make stories about nuclear issues very confusing. The size or weight of a quantity of material is not indicative of how much radioactivity is present. Large quantities of a material may contain only a small amount of radioactivity, while small quantities may contain large amounts of radioactivity. The amount of radioactivity present is usually measured by estimating number of curies (Ci): the more curies, the greater the radioactivity and emitted radiation.

Yet even curies may not always be the best measure to use because, in the case of alpha emitters, much of the radiation may not escape. For radioactive waste a better term is *radiotoxicity*, which is the harmful potential of all these radioactive toxic elements in the spent fuel produced from nuclear power plants. This topic is covered in detail in "Closing the Civilian Nuclear Fuel Cycle," which asserts that certain isotopes of plutonium and the other actinides are the elements that are the most long-lived and hence present the greatest challenge to environmental managers.

References and Other Resources

American Nuclear Society home page. *www.ans.org.*

Centers for Disease Control and Prevention. *Emergency preparedness and response: radiation emergencies. www.bt.cdc.gov/radiation.*

Health Physics Society home page. *hps.org.*

International Atomic Energy Agency. *Factsheets and FAQs: Radiation in everyday life. www.iaea.org/Publications/Factsheets/English/radlife.html.*

National Academy of Sciences. Nuclear and Radiation Studies Board home page. *dels.nas.edu/nrsb.*

National Council on Radiation Protection and Measurements home page. *www.ncrponline.org.*

Nuclear Energy Institute home page. *www.nei.org.*

National Research Council. Committee on Biological Effects of Ionizing Radiation (BEIR V). (1990). Radiogenic cancers. In *Health effects of exposure to low levels of ionizing radiation.* Washington, DC: National Academies of Science. Available at *books.nap.edu/catalog.php?record_id=1224#toc.*

Physicians for Social Responsibility home page. *www.psr.org.*

Radiation Emergency Assistance Center/Training Site (REAC/TS) (part of Oak Ridge Institute for Science and Education) home page. *orise.orau.gov/reacts.*

United Nations Scientific Committee on Effects of Radiation (UNSCEAR) home page. *www.unscear.org.*

(See also state government environmental or public health agencies, radiation control sections.)

Section 2

Nuclear Energy and Other Civilian Uses

Sustainability: Will There Be Enough Uranium and Nuclear Fuel and at What Cost?

Written by Michael R. Greenberg, based in part on an interview with William Szymanski, with comments by Seth Blumsack

Background

The nuclear fuel cycle for energy production begins with exploration for uranium. It ends with transport of fuel, installation, use of the fuel at a nuclear power plant, and removal of the spent fuel to an on-site storage facility. In between are stages that are the focus of this brief and are often overlooked: (1) mining, (2) milling, conversion, and enrichment, and (3) fuel fabrication. (See "Closing the Civilian Nuclear Fuel Cycle," "Nuclear Power Plant Safety Systems," "Decommissioning Nuclear Facilities," "Transportation of Nuclear Waste," and "Nuclear Waste Policy.")

With regard to terms used in this brief, *uranium* refers to mined uranium ore or enriched commercial or weapons grade uranium; *nuclear fuel* refers to commercial-grade enriched uranium that has been processed into fuel pellets and rods for use in commercial nuclear reactors. Another distinction is between fuel fabricating companies and plant operators. Fabricating companies purchase enriched uranium for processing into fuel pellets and rods, that is, nuclear fuel. Operators purchase processed nuclear fuel, not uranium. Companies such as Areva and Westinghouse fabricate fuel and manufacture and service nuclear power plants. But they do not sell to wholesalers and final electricity customers.

Mining

Uranium occurs naturally, consisting primarily of two isotopes, uranium 235 (U-235) and uranium 238 (U-238). (See "Radionuclides.") Both have 92 protons in their nuclei, but U-235 has 143 neutrons compared with 146 for the heavier U-238. U-235 is the most easily fissionable isotope of uranium and is only about 0.7% of natural uranium; U-238 is 99.3%.

There are three major kinds of uranium mining activities. Underground mines are locations where a shaft or tunnel is dug into the rock and workers extract the uranium- laden ore. In open pit mining, the second approach, the overburden (top layer of soil, rock, or other material) is removed and equipment is used to shovel up the rock containing the ore. Both underground and open pit mines use conventional mining methods to first extract ore from rock through drilling and blasting. In situ leaching (ISL) is the third method of extracting uranium from ore for commercial power plants. That method injects chemicals into the ground that oxidize the ore so that uranium can be leached out and placed into solution to be pumped to the surface for processing into uranium concentrate in a processing plant. Because no excavation of rock takes place, ISL operations, when compared with conventional mining and milling, generate significantly less residual waste and cause much less disruption to the landscape. However, ISL is limited to ore contained in porous geological formations through which large volumes of solution can pass.

Milling, Conversion, and Enrichment

Nuclear power plants require uranium fuel containing 3–5% U-235, referred to as "low-enriched uranium" (LEU), hence naturally occurring uranium must be processed and the U-235 portion enriched relative to U-238. The ore from the mine is transported to a mill where it is broken down mechanically and chemically to separate a marketable uranium concentrate product, also called "yellow cake" or uranium oxide (U3O8), from unmarketable minerals. Uranium concentrate in oxide form must be chemically treated to convert it to uranium hexafluoride (UF6), a feedstock for uranium enrichment. The enrichment process uses gaseous diffusion or gas centrifuge to prepare the material for conversion to fuel. The details of enrichment processes are considered proprietary by the commercial fuel producers. (See "Managing the Nuclear Weapons Legacy.")

Fuel Fabrication

From an enrichment plant, the LEU in the form of UF6 is reconverted back to an oxide and made into pellets at a fuel fabrication facility. The fabricator inserts the pellets into fuel rods that are bundled into fuel assemblies for loading into the core of a nuclear reactor. Each stage of the fuel cycle process requires separate business agreements, though some contractors are involved in multiple stages of the fuel cycle. Plant operators must plan ahead for the 12-to-24-month cycle when 20 to 40% of the fuel will need to be replaced. In short, there is nothing simple about the scientific or organizational capacities required to refuel nuclear power plants.

Overall, for context, U.S. Department of Energy experts estimate that 1 short ton (2,000 pounds) of mined uranium yields about 40,000 MWh of electrical power after it is concentrated, enriched, and fabricated.

Alternative Fuel Sources

Some of the commercial nuclear fuel market is supplied from surplus military inventories of natural uranium and LEU derived from blending down highly enriched uranium (HEU) stores. HEU refers to uranium that has been enriched to a U-235 concentration of at least 20%. The first major use of uranium was for nuclear weapons, which created a very specialized "market." Beginning in the early 1990s, uranium was no longer used for weapons production in the United States. Accordingly, stockpiles of uranium declared surplus to military needs became available for commercial use in power plants.

In 1993, the Russian government signed an agreement with the United States, whereby hundreds of metric tons of HEU taken from dismantled Russian nuclear warheads is being blended down to LEU and then sold to U.S. commercial nuclear fuel fabricators for use as reactor fuel. This LEU could supply half of U.S. demand by 2010 but will no longer be available when the agreement ends in 2013 (EIA, 1998). The agreement also has the critical benefit of eliminating weapons grade material.

Another major source of nuclear fuel is inventory purchased by major companies as an investment when prices were lower or as a strategic stockpile held in the event of a significant supply disruption. When Western nations began operating nuclear power plants, the only supplier of uranium in the United States was the U.S. Atomic Energy Commission. The commission supplied the enriched uranium. But the market has evolved into a commercial enterprise with global reach, with enrichment plants in other countries, such as France, Germany, Netherlands, Russia, and the United Kingdom.

What Reporters Need to Know

Reporters should know the basic stages of the nuclear fuel cycle. They also need to know the key issues in today's evolving world market, as well as related health and environmental issues. The next two sections describe five market-related issues and then briefly three major health/environmental issues. The review of health and environment is abbreviated because other briefs focus on these topics.

Five prominent issues drive today's nuclear fuels environment. First, most of the United States supply of uranium comes from abroad. The 104 operating nuclear power plants in the United States use between 50 and 60 million pounds of nuclear fuel a year. U.S. production is between 4 and 5 million pounds a year. In addition to mines outside the United States, a good deal of U.S. supply comes from the drawdown of inventories held by fuel fabricators, suppliers, and governments.

Canada was the number one supplier of ore from major mines located in Saskatchewan. In 2006, Australia supplanted Canada, and now Russia stands second in the U.S. market. Russia, however, does not send uranium concentrate. As previously discussed, Russia sends LEU derived from blended HEU taken from dismantled warheads. Other major sources of raw uranium are Namibia, Kazakhstan, Niger, and Uzbekistan. In 2006, the United States ranked only eighth in ore production.

While the raw resource is available, the capacity to produce fuel sometimes has not kept up. In recent years, nearly 50% of worldwide demand has been met by the drawdown of commercial and government inventories. However, as inventory supply is further drawn down, additional mine capacity will be needed to fill the gap between demand and supply. The industry has not yet made significant investments in new mine capacity to allow for increased production. In fact, world uranium production actually decreased by 5% between 2005 and 2006 (to less than 40,000 tons). The reduced production was in Australia and Canada, where heavy rain and less than anticipated mill performance hindered production. Kazakhstan, Niger, and the United States increased production but could not entirely compensate. This decline appears to be an anomaly related to these conditions. The huge rise in uranium prices over the past several years (see discussion further on) will undoubtedly bring much needed investment into the uranium mining sector. In short, there are good questions to be asked about world, U.S., and perhaps local power plant fuel supply.

A second issue is the development of global competition for uranium. In the short run, there is sufficient uranium ore to meet world needs. But if China, India, France, Japan, and the United States actually expand their reliance on nuclear power plants, new mines and sources will be needed. And there are challenges with development of new mines or with expanding old ones. They require large investment and long periods to prepare environmental and human risk assessments. We should expect as much as 10 years, or even longer, as the time necessary to go from discovery of a major resource to delivery of ore to the market. Old mines can be expanded, but sometimes there are community concerns. In the long term, recycling of spent fuel to power advanced reactors as part of the Global Nuclear Energy Partnership (GNEP) proposed by President George W. Bush in 2006 could generate electricity while stretching out the supply of uranium for many decades.

A related constraint on new and expanded mining is the availability of workers and equipment. The need for other mineral commodities, which also has been expanding worldwide, has drawn resources away from activities to expand uranium ore mining. The dramatic increase in the U3O8 price of uranium concentrate from about $20 per pound at the start of 2004 to around $140 per pound in early 2008 has stimulated the formation of new, often thinly capitalized "junior" companies to engage in exploration. Unlike larger established companies with significant financial resources, the junior companies, if successful in exploration, will require considerable outside investment to take a project to the development stage.

A third important issue is enrichment capacity. In the United States there is only one enrichment facility, located in Paducah, Kentucky. The Paducah plant, operated by United States Enrichment Corporation, Inc. (USEC) and leased from the U.S. Department of Energy, was built in the 1950s. It uses gaseous diffusion technology, which consists of heating uranium hexafluoride (UF6) to form a gas. The gas is then passed through a series of porous membranes. Because U-235 is lighter than U-238, the ever smaller membranes block U-238 and allow U-235 to pass. This technology requires considerable electric power to operate. However, though this is the only facility in the United States, there are a number of other enrichment facilities in the world and that number is growing (e.g., China, France, Germany, Japan, Netherlands).

The plan is to replace the aging and relatively high cost Paducah Gaseous Diffusion Plant operation with more advanced enrichment technology. Not only does a competitive enrichment industry ensure security of domestic supply but also, by being a reliable and competitive supplier of enriched uranium, the United States is assured a leadership role in convincing other nations who may want to enjoy the benefits to forgo building enrichment plants. (See "Closing the Civilian Nuclear Fuel Cycle.")

In June 2006, the U.S. Nuclear Regulatory Commission (NRC) issued a license to Louisiana Energy Services to construct and operate a gas centrifuge uranium enrichment facility in Lea County, New Mexico, known as the National Enrichment Facility. Louisiana Energy Services is a wholly owned subsidiary of Urenco Ltd., a company that owns and operates enrichment plants in Europe. The National Enrichment Facility will use centrifuges to separate the U-235 and U-238 contained in UF6 gas, a process that requires about 10% of the electric power required by gaseous diffusion plants for the equivalent output.

In April 2007 USEC received a license from the NRC to construct and operate what it named the American Centrifuge Plant at the U.S. Department of Energy's Portsmouth site (also the location of a former gaseous diffusion enrichment plant that was shut down in 2001). USEC is to use centrifuges based on improvements to technology that had been developed by the Department of Energy before the government program was terminated in 1985. More recently two other companies have announced plans to build new enrichment facilities in the United States but have not yet submitted license applications to the NRC. General Electric is developing a laser enrichment technology called SILEX for commercial deployment and another company called Areva NC is planning to deploy centrifuges similar to those that would operate at Louisiana Energy Services' National Enrichment Facility.

The fourth issue involves competition for fuel for existing nuclear power programs and the need to ensure access to reactor fuels for developing countries who wish to receive the benefits of nuclear power without having to invest in costly fuel cycle facilities. The United States, France, and Japan are planning to expand their use of nuclear power. China, India, and Russia are relatively new major players in the nuclear power generation business. Competition for nuclear fuel is increasing. For example, the Chinese are talking to African nations and to Australia about uranium and the United States and France about fuel. The Russian nuclear energy industry has been forming joint ventures with major uranium producers in Canada and Kazakhstan. One way to look at these events is to become worried about the competition, that is, the likelihood of global strife regarding uranium supplies. An alternative way is to assume that the market will look at these ambitious plans and expand production. An ambitious expansion of nuclear power should require companies and nations to develop treaties and relationships with each other that can be relatively long lasting and cooperative.

Part of the international effort under GNEP would seek to reduce the use of fossil fuels, guarantee nuclear fuel needs to participating nations and take back spent fuels for reuse, and reduce the spread of proliferation-sensitive enrichment and reprocessing facilities. This effort could involve international arrangements between commercial suppliers and between suppliers and fabricators and plant operators backed by their respective governments. It may also require some level of participation by the International Atomic Energy Agency

(IAEA) in order to provide additional levels of insurance against the disruption of fuel beyond the well functioning commercial marketplace. As long as there is adherence to nonproliferation standards, users would be assured a supply of nuclear fuel even if political disputes arose with supplier nations. A clear issue is the likelihood of successful agreements being abandoned by one side. Setting aside reserves or "banks" of LEU as a further backstop to mitigate supply disruptions has also been proposed. For example, the U.S. secretary of energy announced at the 2005 General Conference of the International Atomic Energy Agency the creation of the "Reliable Fuel Bank" with an initial contribution of 17.4 metric tons of U.S. surplus HEU to be blended down to LEU for an emergency supply for countries not pursuing a nuclear fuel infrastructure. The take back of other countries' spent fuel faces political, regulatory, and public acceptance challenges, but this strategy may be more feasible in the future if and when advanced recycling technologies are successfully demonstrated. There are national and local stories here for reporters seeking to cover the source of local fuel and potential local benefits of replacing existing fossil fuels with nuclear fuels.

Price is the last, but perhaps the most interesting, issue. The United States once was a major uranium producer, but the United States' ore was more costly to extract and use than new sources. Consequently, U.S. production declined substantially to only 2.2 million pounds a year. Production is on the rise because of the increasing global price of uranium. And though the dramatic increase in the per pound U3O8 price of uranium concentrate will lead to growth in U.S. production, the increase is not likely to be large enough to eliminate the need for imports. Nor, we believe, will the current high price necessarily continue, especially as new mines come into operation in countries such as Australia, Canada, and Kazakhstan.

Kidd (2006) characterizes the uranium market as "very immature." His assertion is supported by history. During the late 1960s and early 1970s, U.S. military purchases were decreasing and commercial purchases were increasing. Consequently, yellowcake prices were low, about $7 per pound. The OPEC oil embargo stimulated orders for new nuclear power plants. The price of uranium concentrate rose from $7 per pound at the end of 1973 to $14 in October 1974, to $21 in May 1975, and to $40 in April 1976. The price peaked at $43.40 in May–July 1978 (not adjusted for inflation) (Trade Tech, 2007).

Prices began to plunge, however, as a result of reduced electricity demand associated with the OPEC oil embargo, and events at Three Mile Island reduced the demand for new nuclear power plants. In the wake of the Three Mile Island event, regulators required safety upgrades that added significant delays to plant construction (these were retrofits not anticipated in the initial designs). The combination of lower demand for electric power and construction delays with an attendant rise in costs resulted in the cancellation of many orders for new nuclear power plants. With the cancellation of orders, companies had no choice but to sell on the market at discounted prices the uranium that they had acquired in anticipation of the startup of new reactors. During the mid-1980s, the price was $15–$17 per pound. It fell again during the late 1980s, reaching $7–$8 per pound during 1991 and 1992. Some of this drop can be attributed to the impact of Chernobyl, but it was confounded by countries, including the United States, moving in and out of protectionist trade policy agreements. Each of these policy shifts jarred the nuclear fuel market, making it more unpredictable for investors and driving NUEXCO, the leading uranium trader at the time, into bankruptcy.

In early 2001 the spot price for uranium concentrate began rising. The rise was partly due to the effects of a fire at a uranium processing facility in Australia, a flood at a Canadian mine, and an off-site discharge in the plant in the United States. Thereafter, the price has steadily risen. Yet even in late 2008, the market is adjusting to new circumstances, and the price may fall back but then go back up. Uncertainties affect other stages of the nuclear fuel cycle. Finch, for example, noted in 2006 that the proposed LES centrifuge facility in eastern New Mexico was being challenged for environmental reasons, and that that might delay the opening of that plant, which is expected to supply a good deal of enriched uranium for power plants. The builders of this project are also concerned about finding electrical, aluminum, and other workers, as well as equipment for the project. New findings of high grade ore have been made in Kazakhstan, but the site is relatively inaccessible and will take an enormous investment to develop. In other words, there is no certainty in the world fuel market. Kidd (2006) asserts that it will need substantial investment to meet the world's growing demand. Because plants are likely to operate for many decades, the market should develop to be strong and predictable, but clearly that time has not arrived.

What Are the Risks?

This volume contains more then a dozen briefs that entirely or largely focus on human health risks and one that focuses on ecological risks. This brief will not repeat all of that discussion. Here we concentrate on uranium mining, which is not covered elsewhere in this volume.

Uranium is a heavy dense metal found almost everywhere on the earth at very low noncommercial concentrations. With regard to commercially viable sites, there are three major risk-related issues: radiation risks to workers, environmental contamination, and the production of millions of tons of leftover tailings.

We focus here on worker exposure and mine tailings. Uranium mining can have more long-term effects on worker health than the mining of some other underground resources because the ore emits radon gas. During the 1950s, a good deal of mining was done on Navajo Indian reservations, and a disproportionate number of Navajo workers developed small cell carcinoma of the lung because of exposure to radioactive ore and radon, a natural decay product of uranium. Consequently, it is important to have ventilation in underground mines, and this has been a major issue in the development of new mines. Fortunately, though, uranium mining does not encounter methane gas, which if ignited can cause disastrous explosions that have claimed the lives of coal miners.

Also, there has been a trend away from underground mining. In 1990, over half of the world production came from underground mines; however, by the turn-of-this-century the proportion from underground mines had fallen to one third. In 2006, most U.S. uranium mine production came from ISL operations, and an expanding output from Kazakhstan also comes from ISL. Large underground mines are anticipated to open in Canada over the next several years but these are being planned with state-of-the-art worker protection technology and operator protocols.

As described earlier, the ISL method now prevalent in the United States generates significantly less residual waste and causes much less disruption to the landscape than conventional mining and milling operations. However, the more dominant use of conventional methods in the past has created a substantial amount of residual waste. Any leftover rock, pulverized rock, sludge, and sand from the mining process is called "tailings." (See "Nuclear Waste Policy.") Some of the tailings contain mercury, arsenic, and other toxins. Also, tailings often contain sulfide-bearing minerals, which react with surface or groundwater to create corrosive sulfuric acid that can release other toxins in the tailings and directly kill fish, plants, and other species. In addition, tailings usually contain radioactive residuals, rock that was uneconomical to process. Mines must have a plan to deal with tailings, including a provision to secure a financial bond that would cover the cost for remediation of tailings in the event that the responsible mining company files for bankruptcy. They are regulated by the NRC (10 CFR pt. 40). Appendix A of part 40 covers design and siting of facilities, disposal of tailing waste, decommissioning of structures, monitoring of air, land, and water, and financial responsibilities. Nevertheless, there is a legacy of less than successful performance of tailings management regarding the mines that supplied uranium to the Atomic Energy Commission for military applications. However, under the Uranium Mill Tailing Remedial Act, the Department of Energy is responsible for the legacy environmental management of tailings, including environmental remediation, waste management, and facility transitioning.

Misunderstandings

The problems of Three Mile Island and Chernobyl have meant that an overwhelming amount of attention has been focused on what is wrong with nuclear power. During the past few years, we have begun to see articles about nuclear power that are much more balanced, that is, not necessarily advocating nuclear power but discussing the advantages and disadvantages of nuclear power versus fossil fuels, and acknowledging the track record of the nuclear industry in the United States. While fear of nuclear power continues to exist for some people, the media have helped dispel the myth for others by providing more balanced reporting in recent coverage.

A major shortcoming is that there has been insufficient attention to the parts of the fuel cycle discussed here. The media have been drawn to stories on Yucca Mountain and spent fuel (see "Nuclear Waste Policy"), understandable because they are political and risk-related concerns.

China, India, France, the United States, and other countries have been making financial commitments, sometimes massive, to nuclear power. The decisions by these nations, in contrast to decisions by Austria, Belgium, and some other nations, are by themselves a fascinating and important issue for the media to cover. Also, countries such as Germany and Sweden are reconsidering their policies for phasing out nuclear power as they reconcile the need to reduce carbon emissions with the need to sustain economic growth. The impact of the worldwide financial problems may have a major impact on these investments and is potentially a powerful new story.

With regard to nuclear fuels, and the entire nuclear fuel cycle, the world political and economic implications of international agreements on the mining, processing, selling, using, and reprocessing and storage of nuclear fuels could build positive and ongoing relationships between nations. Of course, friction among nations and private interests could develop as well. That potential merits media attention. Close cooperation between these nations could stand in strong contrast to what happened with international trading of fossil fuels. In turn, successful international cooperation in the nuclear fuel cycle could lead to additional cooperation in developing the hydrogen economy, water desalinization, and other worldwide resource-related issues.

The United States is a major importer of raw materials and feedstocks to produce nuclear fuel. As other countries expand their nuclear power plants, prices should increase in the short run, which would be material for stories when price changes impact customers. But if the expansion is substantial and prices are high, more mining and fuel fabricating will occur, and recycling spent fuel will become a more logical option. The U.S. government's and the nuclear industries' response to the changing global market will, in short, clearly be newsworthy.

References and Other Resources

Energy Information Administration (EIA). (1998). *Surplus defense inventories: A growing source of nuclear fuel for generating electricity.* EIA-98-11. *www.eia.doe.gov/neic/press/press96.html.*

Finch, J. (2006). *Miners and utilities at odds over uranium price forecasts. www.stockinterview.com/News/10022006/Platts1.html.*

Kidd, S. (2006, June). Uranium: The heat continues. *Mining News. www.mining-journal.com/wms_magazine/wms_Mag_Breaking_News.aspx?breaking_news_article_id=3139.*

Trade Tech. (2007). *Uranium as nuclear fuel. www.uranium.info/index.cfm?go=c.page&id=21.*

World Nuclear Association. (2007, May). *World uranium mining.* Nuclear issues briefing paper 41. *www.uic.com.au/nip41.htm.*

Closing the Civilian Nuclear Fuel Cycle and Spent Nuclear Fuel: The Opportunity and the Challenge

Written by Michael R. Greenberg, based in part on an interview with Paul Lisowski, with comments by Thomas H. Isaacs

Background

Federal elected officials, Department of Energy staff, the Nuclear Regulatory Commission, utilities that rely on nuclear power, environmental advocate groups, and other parties with an interest in the energy future of the United States face the difficult challenge of making informed judgments and decisions about the management of spent (used) civilian nuclear fuel. (See "Managing the Nuclear Weapons Legacy.") Traditionally, nuclear fuel, once used in a nuclear power reactor, is referred to as "spent nuclear fuel" in the United States because the United States has not recycled the fuel to recover the unused energy potential remaining. Typically, 1% to 5% of the energy potential of nuclear fuel in a typical light water reactor is used currently, with over 90% of the energy potential theoretically available for recovery and recycle if the fuel is used in a fast reactor. Thus, the phrase "used nuclear fuel" may be more appropriate than "spent nuclear fuel."

One key current issue is whether the U.S. government should pursue reprocessing this spent fuel to extract and recycle the reusable components of the fuels as some select nations are doing. Or should the United States continue its decades-long policy of "once-through" fuel usage where the spent fuel is stored for ultimate permanent disposal, likely to be at least decades away?

This fundamental question, which has been continuously discussed over the past 30 or more years, is now emerging as a key issue along with energy concerns, resource utilization, economic, security, and waste management implications. The challenges and opportunities derive from increasing energy demand, domestic and international politics, and ultimately science and technology. This section summarizes these challenges and ties them to the civilian nuclear fuel cycle.

The United States, the European Union, China, India, Japan, and many other nations of the world with growing economies face the shared challenge of meeting increasing global demand for energy. The U.S. Department of Energy has estimated that world nuclear energy consumption will increase 57% between 2004 and 2030. The largest increases are clearly expected to be in Asia, South America, and Africa, much more than in the developed countries of Europe and North America (EIA, 2006).

In the United States, in addition to an increase in energy demand, there is a growing awareness that we cannot continue our reliance on coal to meet base load electrical generation. One reason is growing evidence that global warming has been occurring largely driven by the emissions from the burning of fossil fuels. (See "Sustainability" and "Global Warming and Fuel Sources.") A second is increasingly stringent fine particulate emission standards. Base load power stations are the workhorses of the electricity system, continuously generating electricity. They halt production only when outages occur and for planned maintenance. In contrast to base load facilities, peak power stations use natural gas and, less often, diesel oil to supply additional electricity during peak demands, typically during the summer when air-conditioning use is high. (Peak load stations typically can start generating electricity in 30 minutes or less, but they are expensive to operate per kilowatt hour produced.) In short, the United States and its counterparts require a reliable source of base load energy production. The great majority of increase in that base load will almost surely be provided by coal or nuclear energy for some time to come. Other options are either intermittent (such as solar and wind), often located far from where the energy is needed (such as with tidal power), or exceedingly expensive until major new technical advances are made (such as with solar electric). We will need and must pursue all effective solutions, but nuclear likely will be among the most important in the coming decades.

Nuclear fuel is one way of meeting the expected global and domestic increase in electricity demand. Worldwide 439 nuclear power plants are operating, another 36 are being constructed, and over 200 are planned (EIA, 2006; World Nuclear Association, 2008). It is plausible that there could be hundreds more new nuclear reactors built during this century to meet the growth in base load electricity needs.

The United States has a "once through" fuel cycle. That is, ore is mined, enriched, and then light water reactor fuel is fabricated. The fuel becomes part of a light water reactor core, and then the spent fuel is recovered, stored on-site until it cools, and finally stored on-site in dry casks. In this concept, only approximately 1% to 5% of the theoretical energy of the fuel is exploited. Recycling would take the spent fuel, separate it into reusable fuel and waste streams, prepare new fuel, and remove the unusable elements for ultimate processing into a permanent waste form.

Nuclear reprocessing was stopped during the presidency of Jimmy Carter largely because of proliferation fears. As noted in a recent report by the International Security Advisory Board (2008), U.S. laws were written in favor of a once-through fuel cycle and against reprocessing because of a fear that spent fuel left in fuel rods would reduce the chances of proliferation. Reprocessing, however, has been practiced in France, the United Kingdom, and Japan. (See "Sustainability" and "Managing the Nuclear Weapons Legacy.") The French, for example, chemically separate elements of spent fuel and create new fuel from the usable component of the reprocessed elements. Recycling extracts and allows consumption of some of the unused uranium as well as the plutonium, which is produced in the original fissioning process. Barron (2007), group executive and chief nuclear officer of Duke Energy, estimates that recycling could provide about one fourth of the U.S. needs.

As mentioned earlier, underlying renewed interest in the recycling of spent civilian nuclear fuel are issues associated with its impact on proliferation and waste management. Both have been major challenges for decades. Separating and converting the material in spent nuclear fuel into nuclear weapons is a major scientific undertaking and financial expense and has never been used by any country that has built weapons. Countries have instead turned to dedicated, often covert, weapons programs. (See "Managing the Nuclear Weapons Legacy.") It is conceivable, however, that with considerable expertise, time, and resources, the reactor plutonium could be diverted to the making of nuclear weapons. Perhaps of greater security concern, the facilities for producing fuel (enrichment plants) or separating the reusable components of the spent fuel (reprocessing plants) could be reconfigured to produce weapons usable materials instead. That having been said, recovering the materials from the spent nuclear fuel and reusing it or rendering it infeasible for use as weapons grade material would be a major international security benefit. Research is continuing on the feasibility of blending impurities into the plutonium that will make it more difficult for the reactor grade plutonium to be used to make nuclear weapons.

Nuclear waste management has also been a fractious issue for decades, epitomized by the ongoing heated debate surrounding the fate of Yucca Mountain. (See "Nuclear Waste Policy.") Currently, spent civilian nuclear fuel is being stored on-site at more than 64 locations. (See "Decommissioning Nuclear Facilities.") While the consensus is that a permanent repository will be needed whether the spent fuel is disposed of directly or reprocessed to reuse the remaining uranium and plutonium, reprocessing might reduce the volume and ultimately the toxicity of the wastes requiring disposal.

In short, in addition to the merits and demerits of using recovered spent fuel in nuclear power plants, the decision to move forward is complicated by its linkage to weapons proliferation and nuclear waste. Strongly held and divergent views continue regarding the impacts of reprocessing and recycling. Even though some anti-nuclear positions appear to have softened, it will be an enormous challenge for proponents of recycling spent civilian nuclear fuel to build a base of support for expanding nuclear power, to trust that recycling will control rather than increase the proliferation of nuclear weapons grade materials and that hazardous residuals can be safely managed. Each of the three represents a major hurdle.

If the proliferation and waste management issues can be addressed, the scientific challenges represent another difficult challenge to and opportunity for proponents of recycling. The operating lifetime of the current generation of reactors is being extended for most U.S. operating reactors, as the Nuclear Regulatory Commission grants license extensions from 40 to 60 years and consideration is being given to developing the scientific and technical understanding to determine whether plants can be reasonably and safely extended ultimately to 80 years of operation. In addition, new standardized reactors are being pursued, using modular designs that can be constructed faster, including passive and redundant safety systems and incorporating enhanced security features. (See "Nuclear Power Plant Safety Systems" and "Protecting Nuclear Power Plants against Terrorism.") Also, the viability of technologies needed to cost-effectively reprocess, fabricate fuel from the recovered uranium or plutonium, and reuse the new fuel as described earlier have not been demonstrated in the United States. Additionally, methods to deal with the waste streams derived from recycling and reuse of spent nuclear fuel must be demonstrated.

What Reporters Need to Know

Political and institutional responses to the issue of recycling civilian nuclear fuel are the logical place for reporters to begin looking for stories. But they must also understand the scientific and technological challenges and the life-cycle economic costs and benefits.

Political and Institutional Information

On the positive side, the potential growth and spread of nuclear power around the world can be instrumental in raising the standard of living and political stability of developing countries. At the same time, it is perhaps the single most important initiative to stem the release of greenhouse gases and the resulting global warming. Significant challenges remain, however, regarding economics, safety, proliferation, waste management, and public acceptance. Perhaps the key institutional and governance issue will revolve around whether we will see the possible emergence of a new nuclear regime or whether the future will evolve in a more "business as usual" manner. Questions about possible regional or international schemes will be timely as initiatives proposed by the United States attempt to craft a future that allows countries that currently have little or no nuclear power to take advantage of the baseline energy potential production in the coming years in a framework that minimizes proliferation and waste management concerns. One of these U.S. initiatives is the Global Nuclear Energy Partnership (GNEP), which grew out of efforts begun in 2006 by the United States to work with France, China, Russia, and Japan to build a global partnership with nuclear fuel supplier nations. The partnership among these original five countries is based on the following principles (for the full text, see GNEP, 2007):

- Expansion of nuclear power to help meet growing energy demand
- Development of advanced technologies for recycling spent nuclear fuel that do not separate plutonium with the goal of eventually eliminating excess stocks of civilian plutonium and drawing down inventories of spent fuel while reducing nuclear wastes
- Development, demonstration, and deployment of advanced reactors that can consume transuranics from the recycled fuel
- Establishment of arrangements to provide reliable fuel supplies and take back spent fuel, to prevent the spread of enrichment and reprocessing technologies
- Development, demonstration, and deployment of advanced proliferation-resistant reactors appropriate for the grids of developing countries
- In cooperation with the IAEA (International Atomic Energy Agency), development of enhanced safeguards to ensure that commercial nuclear power is used only for peaceful purposes

On September 16, 2007, China, France, Japan, Russia, and the United States (all nuclear fuel producing states) joined with Australia, Bulgaria, Ghana, Hungary, Jordan, Kazakhstan, Lithuania, Poland, Romania, Slovenia, and Ukraine and signed the GNEP agreement. Italy, South Korea, Canada, Senegal, and the United Kingdom subsequently joined.

The IAEA likely would also play a major role in GNEP, or another multinational nuclear fuel recycling program. The IAEA has dual functions, to promote the peaceful uses of nuclear power and technology while inspecting and safeguarding facilities and materials to ensure that they are being used only for peaceful purposes. In his famous "Atoms for Peace" speech presented at the United Nations in 1953, President Dwight Eisenhower called for the creation of such an international body. The IAEA was established in 1957 and is headquartered in Vienna, Austria. (See "International Agencies and Policy.") On September 19, 2007, Dale Klein, head of the U.S. Nuclear Regulatory Commission, observed the need for internationally accepted design, safety, and security requirements, including international inspections for a GNEP-like program (see Kerr and Pomper, 2005).

The domestic reaction to GNEP has been mixed. Beginning on November 14, 2007, the U.S. Senate Energy and Natural Resources Committee began hearings on GNEP. Ling (2007a, 2007b) reports that the White House requested $405 million for fiscal 2008, the Senate approved $243 million, and the House of Representatives $120 million. A total of $181 million was approved.

Noting that support for GNEP was shrinking, Ling (2007a, 2007b) describes elements of the hearing. For example, Senator Peter Domenici (D-NM) criticized the proposed 50-year program as longer than the United States can wait, especially because of the questionable fate of the Yucca Mountain repository. Second, Peter Orszag, director of the Congressional Budget Office, testified that the life cycle cost of recycling would be 25% more than direct disposal, which one of our reviewers noted is estimated to add 18 cents per month to the average consumer's bill. Finally, the hearing considered a National Research Council report (2007) that criticizes GNEP as too aggressive and suggests a more research-oriented effort, which, in fact, was what the Department of Energy indicated that it was supporting.

While not part of the congressional hearings, Barron (2007) has raised one additional domestic institutional challenge that is as fundamental as any discussed at the congressional hearing. Pointing to both the rapid change of leadership within the Department of Energy and a lack of consistent funding, he recommends a government-chartered corporation (like the Tennessee Valley Authority) that would recruit its leadership and draw funds from electricity generators and customers rather than rely on annual congressional budgets. The essence of his request is to build an organization that is stable, both in leadership and in funding, to guide this important challenge and opportunity.

Balanced against these multiple concerns is that GNEP could help improve the standard of living and political stability of developing countries while helping reduce greenhouse gas emissions.

Local stories will be a little more difficult to find here, but they can be found. The reporter can focus on local nuclear power plants: the amount of recycled nuclear fuel compared with the fuel equivalent of coal measured as number of coal-carrying trains, greenhouse gas emissions avoided, safety issues related to the blended fuel compared with those associated with other fuels, transportation of nuclear fuel and waste products, and job and other local economic benefits and costs. There may be some direct impact on the local community. For example, on December 22, 2007, the *Paducah Sun* reported that a proposed nuclear fuel recycling plant that the city had sought was on hold because the Department of Energy did not include it in its environmental impact assessment. The mayor of Paducah noted that they would turn their attention to keeping the existing Paducah Gaseous Diffusion Plant operating as long as possible.

Science and Engineering Challenges

Safeguarding existing reactors is an ongoing priority. Also, new reactors will be needed. Some nations that want to participate in a GNEP-like program (e.g., Bahrain, Ghana, Jordan, Lithuania, Poland, Vietnam) have small electricity delivery grids that cannot host a standard nuclear power plant. Hence, new smaller scale reactors (10–350 MWe) must be developed as well as grid systems that can accommodate them. Every one of these scientific challenges is an opportunity to build positive political and economic relationships among nations. (See "Nuclear Power Plant Safety Systems," "Three Mile Island and Chernobyl," "Monitoring and Surveillance of Nuclear Waste Sites," "Long-Term Surveillance and Maintenance," and "Managing the Nuclear Weapons Legacy.")

The United States has not reprocessed and reused nuclear fuel for over three decades, and it needs to be able to demonstrate that it can recycle civilian nuclear fuel without separating plutonium and in a manner that clearly shows that such recycling will not aggravate proliferation concerns. It also must develop it in a way that can compete economically in the marketplace of the future. These are important but difficult challenges for the scientific and technical community, even if adequate resources are provided.

An important technological challenge is the task of the "fast reactors." Briefly, the United States' fleet of light water nuclear reactors uses water to cool fuel elements and transfer the heat to make electricity. When neutrons collide with water, they slow down, which means that they cannot effectively fission transuranic elements, most notably plutonium. (See "Managing the Nuclear Weapons Legacy.") Transuranic elements are man-made radionuclides and among the heaviest with an atomic number greater than uranium, such as some isotopes of americium, plutonium, neptunium, and californium. Fast reactors use liquid metals, such as sodium and lead, to cool the fuel. When neutrons collide with sodium they slow down much less than when they collide with water. Fast reactors can fission many transuranics that will not fission in a light water reactor. In other words, fast reactors can use the plutonium as fuel, extracting much more energy per unit of original uranium fuel, while simultaneously drawing down the inventories of plutonium that would otherwise be a potential material to be diverted into nuclear weapons. It could conceivably also reduce the inventory of wastes that would require long-term storage and ultimate disposal in a repository. (See "Nuclear Waste Policy.")

Summarizing with regard to science and technology, reporters will want to follow the development, testing, and scaling up of technologies that are intended to reuse civilian nuclear fuel and consume potentially weapons grade material, while protecting workers and others who potentially could be exposed. These technologies may be developed in the United States, but there is also a strong likelihood that the technology will be developed, tested, and initially implemented in France, Japan, and Russia, although U.S. corporations may be involved.

Economic Challenges

Developing and operating fast reactors is mainly an economic challenge. If the industry expands and the technology advances along with the expansion, presumably the economic challenge will become easier to forecast.

Opponents of the effort argue that the aggressive pursuit of the technology implies a multi-decade government-subsidized nuclear reprocessing program that is not required. Currently, one might estimate that generating electricity from a breeder reactor might cost 20–30% more than light water reactors. This estimate is quite uncertain, however, since no commercial plants have been built in the United States and few breeder demonstrations have been built anywhere. But as prototypes are developed and commercialized, the cost may come closer to the cost of light water reactors.

Will the need for advanced breeder reactor nuclear technology increase during the 21st century? Uranium is currently reasonably plentiful, though prices have been rising. Depending on the speed and degree with which nations turn to nuclear power, likely sometime during this century, uranium could become expensive enough that enhanced utilization through recycling of unused fuel could become economic. For example, in the late 1990s, a pound of natural uranium cost $10. The spot market rose to over $100 per pound in 2007, though recently it has settled a bit from that high to about $60 to several of our experts. Accordingly, one type of fast reactor technology challenge that might be desirable is the possibility of breeder reactor technology. Breeders can recover more than one half of the potential energy in fuel; some sources say up to 75%. A breeder reactor is a fast fission reactor that uses neutrons to create new fuel. For example, it converts stable uranium-238 into plutonium-239. The "fast breeder" actually produces more fuel than it consumes because of the conversion of the non-fissionable uranium-238 (which constituters more than 99% of uranium reserves) into fissionable plutonium-239 at a faster rate than the original fissionable fuel is consumed. Different breeder reactors have been proposed based on sodium, lead, and gas cooling. In addition to economic drivers, countries might pursue breeder technology for energy security reasons as well. Japan, significantly, has developed the entire fuel cycle; as a country without substantial intrinsic energy resources, having its own capability to reprocess and recycle will provide much greater assurance of access to needed energy in the coming decades. India is considering a breeder reactor to use its ample thorium supply, another raw material that can be converted into a fissionable element, in this case, uranium-233. India possesses about one third of the world's thorium, compared to less than 1% of the world's uranium. (See "Sustainability.")

Ironically, the first electricity ever produced by nuclear power came from a small fast breeder reactor, the Experimental Breeder Reactor-I (EBR-I) in 1951. However, breeder reactors did not make up a significant part of the expansion of nuclear power. Instead, light water reactors, developed from the technology used to power nuclear submarines, became the reactor of choice for the first generation of plants around the world. The breeder reactor core consists of thousands of stainless steel tubes containing a mixture or plutonium and uranium oxides, including about 15–20% fissionable plutonium-239. The core is surrounded by the "breeder blanket," which consists of tubes containing uranium oxide. The heat from the fission produced in the core heats up the surrounding sodium, which heats water to produce steam for electricity. The reactor can be designed to create perhaps 20% more nuclear fuel than it consumes, which potentially means substantial economic benefits in terms of fuel costs saved and waste disposal avoided.

Summarizing, political and institutional arrangements, scientific and technological challenges, and life-cycle economic costs compared with benefits are at the heart of the recycling civilian nuclear fuel issue, and the breeder reactor challenge may or may not become part of it. Reporters are likely to be writing stories about these at the global, national, and sometimes local levels for decades.

Misunderstandings

Perhaps the biggest misunderstanding that arises with the back end of the fuel cycle involves the implications of the choice between reprocessing the used fuel and disposing of it directly in a geologic repository. This is in some senses a false choice. Whether or not one separates the unused uranium and produced plutonium for recycling in a new fuel load, wastes remain that are potentially hazardous for very long periods (perhaps hundreds of thousands of years) and need to be essentially permanently isolated from the "accessible environment." A geologic repository can do this in either case. What reprocessing and recycling can potentially do is unlock the full energy potential of the uranium and plutonium and draw down the inventories of these materials through recycling, easing, though not eliminating, proliferation concerns at the same time. It may also reduce the volumes and radiotoxicity (discussed later) of the resulting wastes, though this remains an issue to be pursued further. There is some prospect that advanced reactors and their associated fuel cycles decades from now may be able to begin to significantly eliminate many of the more difficult waste elements. But this is far off and subject to a great deal of R&D before the practicality and infrastructure requirements can be determined.

Another possible misunderstanding relates to the number of repositories that might be needed if there is a significant expansion in future U.S. nuclear power production. There is currently a legislated limit on the amount of waste that can be disposed of in Yucca Mountain, but this is a (legitimate) political limit, not a technical one. The limit (currently 70,000 metric tons of heavy metal) was set at approximately half of the amount of spent fuel that was expected from the first generation of U.S. nuclear plants when the original nuclear waste policy act was passed in 1982. The limit applies only until a second repository becomes operational and was intended to provide a degree of geographic equity since 80% of nuclear power is generated in the eastern U.S. and there was an expectation that the first repository might be sited in the West. Should the United States successfully site a second repository, the limit on Yucca Mountain would no longer apply and there are many ways to increase the capacity of these repositories. Another alternative, should Yucca Mountain succeed, would be to change the provision in the law. Though it is too early to tell, regional or international schemes might eventually limit the needed number of repositories world wide as well; however, dividing the amount of projected waste by 70,000 metric tons is not a reliable or accurate way to determine the number of required repositories.

Pitfalls

Perhaps the greatest potential pitfall, assuming proper attention to safety, economics, the environment, and security is public and political acceptance. Both the reprocessing of used fuel and the siting of geologic repositories have suffered significant setbacks in many countries. The siting of a repository is particularly difficult. Public trust is often lacking and political considerations make it very difficult for elected officials to tolerate even consideration of such a facility in their domain. There are no easy or sure answers to this problem where democratic forces are in play, especially in countries that have strong state governments or their equivalent. Local communities often see the potential benefits of such a project as a tradeoff worth considering if they are treated with respect and as a full partner. State-level governments tend to be located elsewhere, because repositories are often sited in relatively remote locations, and view siting in their domain as a sign of weakness or stigma, potentially risking political defeat. Nonetheless, there are examples of countries where the repository programs appear to be succeeding and others where changes in approach may change their fortune. Finland and Sweden are the best examples of the former and France and Canada may be examples of the latter, though that still remains to be seen. As people begin to see the emerging links among energy sufficiency, environmental integrity, global warming, national security, economics, and the back end of the fuel cycle including disposal, the prospects may be better for success on this important but challenging issue.

While the above are the key issues that lead to misunderstandings, we have elaborated upon others that are related to these but are more challenging to capture in stories. With regard to recycling spent nuclear fuel, an important pitfall is the use of the word "curie" to imply hazard and risk. A curie is a symbol of radioactivity named after Marie and Pierre Curie. Curies and becquerels measure the number of radioactive atoms disintegrating per second. But disintegrations do not translate directly into a measure of hazard, nor do they imply anything about exposure and risk. The media should focus more on the concept of "radiotoxicity." The critical elements of spent fuel are uranium, plutonium, and other actinides. These radionuclides emit alpha, beta and gamma radiation. Their decay periods can be thousands or hundreds of thousands of years. The radiotoxicity of spent fuel is the weighted sum of all these toxic elements in the spent fuel. Radiotoxicity is not a measure of risk. It is a measure of the cumulative total amount of toxic material that must be managed. Radiotoxicity is important because plutonium, by far, is the largest contributor to the long-term radiotoxic inventory of spent fuel. The spent fuel fission products, such as iodine, and to a lesser extent cesium and technetium-99, have relatively short toxic lives. (See "Radionuclides and Human Health Effects.")

The minor actinides, such as americium and neptunium, make a larger contribution to long-term radiotoxicity than fission products. But their radiotoxicity is 10 times less than that of plutonium. This means that after 300 years, plutonium residuals account for about 80% of the radiotoxic inventory of spent fuel, and this proportion steadily increases with time. Hence, when talking about management of spent fuel, it is important to eliminate actinides, especially plutonium, not only because of its potential as a weapon but also because of its long-term radiotoxicity.

Recycling would reduce exposure to the most hazardous radionuclides by capturing, reusing and consuming the plutonium and other actinides that have the highest radiotoxicity. A comparison of recycling with the Yucca Mountain facility is instructive. If Yucca Mountain is opened and waste is stored there, some of the materials will be radioactive for 300,000 years, which presents an unprecedented institutional challenge: much of the waste will be a hazard for as much as 25,000 years. Will society in the future be able to read a "keep out" sign as we approach 25,000 years? Recycling would consume these elements, leaving some fission products that would need management for 300 years, which seems plausible to accomplish.

The French currently separate spent nuclear fuel, 15% of which is used in light water reactors. This step reduces the amount of waste by a factor of five (Barron, 2007, suggests a factor of 8). The process drastically reduces radiotoxicity from hundreds of thousands of years to 9,000 years. That is still a long time, but the vitrified glass logs placed in stainless steel boron containers are certified for 10,000 years. (See "Nuclear Waste Policy.") Overall, media coverage of waste management needs to focus more on radiotoxicity rather than curies and becquerels.

With regard to pitfalls that directly relate to recycling of civilian spent nuclear fuel, the cost issues have been neglected, perhaps because of uncertainty with all the options. The current cost estimate for the Yucca Mountain repository is $54 billion; this cost will clearly exceed $60 billion. If the number of reactors increases, a second tunnel would be needed, and the costs of that will substantially increase the overall costs. This is not to say that recycling spent nuclear fuel will eliminate the need for a repository or interim storage sites. What it might accomplish is relieving the pressure for an immediate decision, which might allow scientists and engineers to develop safer and more economical solutions. Despite a high degree of cost uncertainty, there needs to be more coverage of costs, including the reasons for the high level of uncertainty. Without additional coverage, proponents and opponents alike can too easily assert certainty and advocate based on misleading cost estimates.

Another pitfall is accepting the assertion that the public will not agree to new nuclear-related facilities. (See "Public Perception.") The literature suggests that the majority do not want new facilities near them. However, the "not in my backyard" response is not universal. Some areas with existing facilities appear to be open to discussion of the possibility. In an industry-funded study, Bisconti Research, Inc. (2007) reported on a national survey of people who live within 10 miles of one of the existing 64 nuclear power plant sites that host all the nuclear power plants in the United States. Seventy-one percent agreed that nuclear waste can be stored safely at the plant sites until moved to a permanent disposal facility. With regard to recycling, a recent MIT study (Ansolabehere, 2007) found that only 28% of respondents believe that nuclear waste could be stored for long periods. After being informed about recycling of spent fuel, 60% said they supported the expansion of the Department of Energy's recycling program. Neither of these studies is definitive. The point is, however, that reporters can learn a great deal about local preferences by speaking with elected officials, other responsible government experts, and local residents.

The media should cover the economic implications of federal government decisions regarding recycling of spent civilian nuclear fuel. While some do believe that we can continue not to recycle under existing federal policy, others believe that continuing the no reprocessing policy in the United States will kill the U.S. nuclear industry, creating serious economic and political consequences for the United States because France, Japan, South Korea, China, Russia, India, among others, are moving swiftly toward adopting nuclear power. What will be the position of the United States globally if it does not play a major role? After many years of being the most advanced technological nation in the world, what will happen to the United States if it does not participate in a multinational nuclear power renaissance?

A final pitfall is that there is too much focus on single components of the nuclear fuel cycle issue, such as waste management and implications for proliferation. Not enough focus is placed on managing the entire fuel cycle. For example, on one hand, one 2,500 metric ton nuclear fuel recovery facility would provide the same energy equivalent of what comes from the Alaska oil pipeline. That fact implies support for recycling of spent nuclear fuel. On the other hand, the challenge of developing, building, and operating a set of new technologies that would follow a major commitment to recycling spent nuclear fuel would be a challenge to at least match the construction of the Alaskan pipeline, which was a monumental effort. In other words, we could use more articles that probe the comparative advantages and disadvantages of our energy options, not articles that overlook the big picture by focusing on only a single piece of the puzzle, even though each piece of the puzzle is important.

References and Other Resources

Ansolabehere, S. (2007, June). *Public attitudes toward America's energy options: Insights for nuclear energy.* MIT-NES-TR-08. Washington, DC: U.S. Government Printing Office.

Barron, B. (2007, July 17). Comments made at National Association of Regulatory Utility Commissioners meeting, Division of Nuclear Waste Management. *www.narucmeetings.org/Presentations/DiscussionOfNuclearWasteMgmt_DukeEnergy.pdf.*

Bisconti Research, Inc. (2007, July–August). *National survey of nuclear power plant communities.* Conducted for the Nuclear Energy Institute. *www.nei.org/filefolder/brief_summary_10-mile_radius_national_survey_07.doc.*

Energy Information Administration (EIA). (2006, May–June). *International Energy Annual, 2004. tonto.eia.doe.gov/reports/reportsD.asp?type=International.*

Global Nuclear Energy Partnership (GNEP). (2007). *Global Nuclear Energy Partnership statement of principles. www.gneppartnership.org/docs/GNEP_SOP.pdf.*

International Security Advisory Board. (2008, April). *Report on the proliferation implications of the global expansion of civilian nuclear power. www.state.gov/documents/organization/105587.pdf.*

Kerr, P., & Pomper, M. (2005, March). *Tackling the nuclear dilemma: An interview with IAEA Director-General Mohamed ElBaradei.* Arms Control Association. *www.armscontrol.org/act/2005_03/ElBaradei.asp.*

Ling, K. (2007a, November 13). Nuclear waste: Pair of Senate hearings to examine future, legacy of waste. *Environment and Energy Daily.*

Ling, K. (2007b, November 15). GNEP losing support from all sides. *Environment and Energy Daily.*

National Research Council. Committee on Review of DOE's Nuclear Energy Research and Development Program. (2007). *Review of DOE's Nuclear Energy Research and Development Program. www.nap.edu/catalog.php?record_id=11998.*

U.S. Department of Energy. (2006, May). Report to Congress. *Spent nuclear fuel recycling program plan.* Washington, DC: U.S. Government Printing Office.

U.S. Department of Energy. (2006, July). *Advanced fuel cycle initiative (AFCI) comparison report, FY 2006 update.* Washington, DC: U.S. Government Printing Office.

World Nuclear Association. (2008, May 23). *World nuclear power reactors 2006–08 and uranium requirements. www.world-nuclear.org/info/reactors.htm.*

Nuclear Power Plant Safety Systems

Written by Michael R. Greenberg, based in part on an interview with B. John Garrick, with comments by Chris Whipple

Background

Over 400 licensed nuclear power plants operate in the world, including more than 100 in the United States. This brief is about nuclear power plant safety—in reference to both the technology and the people who operate it. We focus on efforts made by the nuclear industry and government to maintain and increase the improved safety and performance of nuclear power plants. More specifically, this brief addresses independent redundant safety systems, defense in depth, and trends toward inherently safe reactor systems. Other safety-related briefs review transportation, lessons learned from the TMI and Chernobyl events, decommissioning, closing the civilian nuclear fuel cycle, protecting plants against terrorism, and nonproliferation.

Identifying the Issues

We need to be assured that designers, builders, operators, plant managers, and postclosure staff are up to the challenge of designing, operating within the regulations of the Nuclear Regulatory Commission (NRC) regulations, protecting plant security, maintaining reliable safe operations, managing, and eventually decommissioning the facility. (For further discussion of these issues, see, for example, Weick and Sutcliffe, 2001.) Nuclear power plants are engineered to be as inherently safe as practicable, backed up with multiple protective barriers. For example, nuclear fuel pellets are enclosed in corrosion resistant cladding composed of zirconium. The reactor primary cooling system acts as another containment system. And the reactor core and primary cooling system are enclosed in a containment building designed to withstand high pressures and prevent the release of radioactive materials in the event of a postulated accident. Nuclear power plants also have secondary physical structures to protect the radioactive core. Other engineered systems include automatic control devices in the event of operator error, backup systems if the initial operating system fails, and operations and technology that permit recovery from equipment malfunctions. Outside the reactor containment building, every commercial nuclear plant is surrounded by an "exclusion" area in which no development is permitted. Finally, in the United States, all nuclear power plants are required to have NRC-approved emergency preparedness plans, an on-site emergency operations center, and an off-site emergency center with facilities for media personnel. These engineered systems and legal requirements and guidelines must work and be trusted by those who rely on them. (For descriptions of the role of the Institute of Nuclear Power Operations, the Nuclear Energy Institute, the International Atomic Energy Agency, and the World Association of Nuclear Operators in developing guidelines and procedures for the nuclear industry, see the U.S. Department of Energy and NRC Web sites listed at the end of this essay.)

While invariably the power plant is the focal point of media attention with regard to safety, issues that flow from the use of nuclear power plants follow the nuclear fuel cycle both upstream and downstream, including concerns about worker exposure, air and water pollution resulting from mining and production of the fuel, transportation of the fuel to the power reactor sites, and storage and movement of waste products. Each of these is an issue for media coverage. (See "Radionuclides and Human Health Effects," "Sustainability," "Transportation of Nuclear Waste," and "Nuclear Waste Policy.")

After safety, the other important issues are economic and environmental implications of nuclear power. How will the market allocate shares to fossil fuels, solar, wind, conservation, and other options? We expect substantial regional differences within the United States that will be good opportunities for stories. There is growing concern about greenhouse gases resulting from fossil fuel use. What are the environmental consequences of substituting nuclear fuel for fossil fuels, especially coal? If a nuclear plant were built in the journalist's area, what is the equivalent in tons of fossil fuels not consumed? How significantly would nuclear power decrease greenhouse gases? Each of these is a good issue for journalists to explore.

What Reporters Need to Know

Reporters who chose to focus on power plants should determine the type of power plant that exists or is proposed for the area. Is it a boiling water reactor? Or is it a pressurized water reactor? Who manufactured it? What are its safety systems? How old is it? What is its safety record? Have there been any unanticipated events? How were these addressed? How much down time has the plant had compared with operating time? How does its operating time compare with the norm for these kinds of reactors? What explains the downtime? Does the plant use risk assessment to evaluate these and other questions? How has it used risk assessment to improve safety and operations?

When the reactor shuts down, fission stops, but how long will decay heat produced by fission products require operating cooling systems? With regard to waste management, depending on the type of reactor, plants are typically refueled every 18 months. Operators take out the fuel elements that have been used the most and place them in a pool, where decay heat is removed. These waste management structures (i.e., the spent fuel cooling pool) are in the containment building. A small plane could not penetrate the building. Nor could a typical large passenger airplane. A high performance airplane could possibly cause a problem. A study completed by Electric Power Research Institute (EPRI, 2005) estimates that a terrorist attack from an airplane crash would not penetrate the containment and damage the spent fuel cooling pool. Even though there have been tests performed to demonstrate containment capability, there is no experience with such an event. After the used fuel assemblies have cooled sufficiently, they can be moved to a licensed, secure on-site dry storage pad on the utility site. (See "Nuclear Waste Policy.") Reporters can ask questions about these waste management systems and operations used to manage spent fuel (e.g., used fuel), including any transportation to and from the plant. Worker safety education, training, monitoring and record-keeping protocols are possible follow-up issues as well.

With regard to nuclear power plants, risk assessment is a series of steps that answers three questions. What can go wrong? How likely is it that it will go wrong? What are the consequences? In other words, risk assessment analyzes possible accidents along with a distribution of consequences produced by interacting events that could lead to a radioactive release.

Finally, a study would include an analysis of likely responses to such an event, including emergency response, evacuation and sheltering in place, and prevention of the problem from occurring again.

With regard to nuclear power plant safety, risk assessment addresses the likelihood (probability) and consequences of postulated incidents involving hazardous materials, including any transportation of fresh fuel to the plant and used fuel from the plant. Worker safety, training, radiation monitoring, incident reports, and health records of employees are possible follow-up issues as well. Illustrations would include the probability of release of hazardous radioactive materials, the maximum concentration of those materials in the facilities, the maximum radiation exposure to workers in and near those facilities, the capability of design features and operators to prevent its release from the containment, the diffusion of these materials into the environment in the event of a release, and the exposures to the general public that could result from release of hazardous radioactive materials from containment. Quantitative risk assessment (QRA) or probabilistic risk assessment (PRA) are the terms used by the nuclear industry. For purposes of this brief, they should be considered to be the same. Nuclear industry guidelines assume that power plants will use a PRA, and the NRC requires a PRA for all licensed nuclear power plants. Enforcement action is taken if NRC inspectors find the PRA to be out-of-date. Indeed, the idea for risk assessments for nuclear power plants started in the mid-1950s. In 1957, the U.S. Atomic Energy Commission's report "Theoretical Possibilities and Consequences of Major Accidents in Large Nuclear Power Plants" (AEC, 1957) estimated the likelihood and consequences of an accident. Studies such as these made it clear that more information and a method of assessing risk more quantitatively was needed to improve safety. In 1975, a reactor safety study by the MIT professor Norman Rasmussen (NRC, 1975) and later studies of the Zion, Illinois, and Indian Point, New York, plants focused on reducing the risk with specific safety features. By the 1980s, QRA had become the industry-standard used to identify vulnerabilities in plant designs and operating procedures, so that significant contributors to accident risk could be mitigated.

Another likely interesting line of questioning for a reporter involves how safety practices in the nuclear power plant compare with practices used in nearby refineries, chemical plants, and other sites that handle hazardous materials. Readers should understand the safety requirements of the nuclear power industry compared with those of other industries that deal with hazardous materials. Reporters can ask about hardening of structures, backup systems, vehicles that move hazardous substances, and worker training and performance requirements. They can inquire about the use of risk assessment as a process to improve safety and performance. Finally, nuclear power plants are required to have evacuation plans. Ask about these at the local nuclear plant, and ask about similar plans at refinery, chemical, and other facilities that handle hazardous materials. Do these exist on paper? How often are they updated? How do they work, and are they tested? How often? This is not to say that chemical plants do not have good safety programs. It is to say that looking at other potential hazardous facilities can provide a useful context.

What Are the Risks?

The much greater number of licensed power reactors in the world operating now compared with 30 years ago increases the opportunity for an accidental release or deliberate misuse of nuclear materials. It also increases the possibility of diversion of spent fuel and other unacceptable uses of radionuclides. (See "Nuclear Nonproliferation" and "Dirty Bombs.") There is no denying these possibilities, but security is far tighter than ever before, so the overall risk is less.

Also, the plants are working better in the United States, Japan, France, and some other countries because of experience and almost continual analysis, evaluation, and updating. Worldwide, the industry has accumulated approximately 10,000 plant-years of operating experience. It has become a leader in monitoring, surveillance, and worker training. Plant-specific risk assessments became essential after the Three Mile Island and Chernobyl accidents, and after the risk assessments were done on several nuclear power plants). Having established risk assessment as a design and planning tool, management and regulators use QRA to guide development, testing, and operation of a new generation of nuclear power plants under consideration. Despite all the efforts, learning from accidents and QRAs, no one can guarantee that there will not be any new accidents.

The challenge to local communities, to state and federal government, and to energy-dependent business is how to assess the risks and benefits of increasing our use of nuclear power. Undeniably, the United States and some other countries need an alternative source of energy. How can our society reasonably balance the strong opinions about the subject of our energy future? Can we secure more informed public involvement than we currently have? This is a daunting challenge for the next decade, and the role of the media is clearly important in increasing informed dialogue.

Nuclear power plants are being designed and operated to follow the principle of independent and separate safety trains. Plants contain independent power sources, pumps, and other vital equipment protected and hardened in locations that have been located to decrease their vulnerability. A submarine with independent and sealed rooms is a good analogy for today's nuclear power plants. If something were to go wrong in one room, the others are protected and backed up. Think of nuclear power plant workers being trained the way astronauts are trained. The Japanese and French developed—and the United States adopted—the idea of using simulators and information centers to improve and sustain the quality of the work force. Designers and operators of key systems can retrain, test new ideas and approaches, and practice on a mock up of an actual facility.

The ultimate goal of designers is to build a new generation of nuclear power plants that are "inherently" safe. This means designing the plant so that it has no need for off-site electrical power and cannot have a loss of coolant failure or experience "transient" events, in other words, the plant stays in a safe configuration where cooling is maintained and where radionuclides cannot move. These steps would reduce risk and increase the efficiency of operation, leading to increased economic benefit of reactor operations.

Development of these new reactor plans will take 20 to 40 years and will require a highly educated and talented work force. Nuclear engineers are already among the highest paid engineers. However, in the United States, the work force has declined, qualified nuclear engineers are scarce, and university departments of nuclear engineering have been discontinued or reduced in size. And university reactors to train the next generation of nuclear engineers have been closed. If nuclear power is going to play a larger role in the United States, there needs to be a rebirth of nuclear engineering university-based programs, and the NRC and the U.S. Department of Energy have developed and implemented such an effort.

Misunderstandings

The track record of the industry should be allowed to speak for itself rather than continuously being tethered to the Chernobyl and Three Mile Island accidents. (See "Three Mile Island and Chernobyl.") There have been two serious reactor accidents in 10,000 plant-years of operating experience. The Chernobyl accident occurred in a reactor type that is not like any used in the United States, and one that could not be licensed in the United States. At Three Mile Island, a series of design shortcomings that were compounded by human errors caused the meltdown. These have been deeply examined and lessons learned have been applied to avoid a reoccurrence. What happened on these two occasions cannot be negated, but it can be contrasted with what occurs today and is planned for the future.

Pitfalls

In recent years the coverage of nuclear power has changed. It appears to be more balanced among proponents and opponents. The likelihood that nuclear power is making a comeback is being covered, and there is coverage of those who formerly were opponents but now are more open to the idea. Political vulnerability of fossil fuels, higher costs, greenhouse gases, and other externalities tell us that the combustion process cannot be the way of the future. Nuclear power can become part of a solution during the next 10 to 30 years. It can be the technology that provides the energy for the hydrogen fuel cycle, which is a system of providing hydrogen fuel for automobiles and other energy-dependent parts of the economy. Nuclear power can generate enough heat to desalinate water or to produce liquid fuel for transportation from coal, both of which are likely to become more important in the United States. Reporters should not abandon their critical surveillance of the nuclear power industry, but they should cover the potential benefits as well as costs, and they should present information on the safety record of their local facilities more often than they do now.

References and Other Resources

Electric Power Research Institute (EPRI). (2005, December). *Nuclear power plant risk analysis and management for critical asset protection (RAMCAP): Trial applications summary report.* No.1011767. *mydocs.epri.com/docs/public/000000000001011767.pdf.*

Garrick, B. J. (2004). Nuclear power: Risk analysis. In C. Cleveland (Ed.), *Encyclopedia of energy* (Vol. 4, pp. 421–31). Washington, DC: Environmental Information Coalition, National Council for Science and the Environment.

McFarlane, H. (2006, May 24). *Advanced reactor technologies.* Paper presented at the meeting of the California Council on Science and Technology, Sacramento.

U.S. Atomic Energy Commission (AEC). (1957, March). *Theoretical possibilities and consequences of major accidents in large nuclear power plants.* Report WASH-740. Washington, DC: U.S. Government Printing Office.

U.S. Department of Energy home page. *www.energy.gov.*

U.S. Nuclear Regulatory Commission home page. *www.nrc.gov.*

U.S. Nuclear Regulatory Commission (NRC). (1975, March). *Reactor safety study: An assessment of accident risks in U.S. commercial nuclear power plants.* Report WASH-1400 (NUREG-75/014). Washington, DC: U.S. Government Printing Office.

U.S. Nuclear Regulatory Commission (NRC). Division of Systems Research. Office of Nuclear Regulatory Research. (1990). *Severe accident risks: An assessment for five U.S. nuclear power plants.* NUREG-1150. Washington, DC: U.S. Government Printing Office.

Weick, K. E., & Sutcliffe, K. M. (2001). *Managing the unexpected: Assuring high performance in an age of complexity.* San Francisco: Jossey-Bass.

Three Mile Island and Chernobyl: What Happened and Lessons Learned

Written by Michael R. Greenberg, based on an interview
with Robert J. Budnitz, with comments by J. Samuel Walker
and Buzz Savage

Background

In 1979 and in 1986, many people watched and read about the accidents at the Three Mile Island and the Chernobyl nuclear power plants, respectively. Almost three decades later, after the publication of numerous technical reports and opinion pieces about these momentous accidents, reporters asked us to revisit the accidents, summarizing what happened and what lessons have been learned. I interviewed Robert J. Budnitz, a central figure in both accidents and asked him to reflect on Three Mile Island and Chernobyl, and J. Samuel Walker, author of many books about nuclear issues, critiqued the brief, as did Buzz Savage, a senior U.S. Department of Energy scientist.

Robert J. Budnitz describes himself as a safety expert. He received a B.A. from Yale and Ph.D. in physics from Harvard. Budnitz spent much of his career as an adviser/consultant. When I interviewed him, he was on his way to Lithuania, where he is advising the government about its Soviet-era nuclear power plants. In the United States, he is an employee of Lawrence Berkeley National Laboratory of the University of California at Berkeley.

Most important from the perspective of this brief, Budnitz was the deputy director of research and later the director of research of the U.S. Nuclear Regulatory Commission (NRC) from 1978 to 1980. The Three Mile Island event occurred on March 28, 1979, 6 months into his tenure at the NRC. The NRC charged him to serve as the technical coordinator of an NRC special-inquiry team to determine what happened at TMI and why it happened, that is, in the aftermath of the accident. The inquiry group, which included 65 NRC staff, conducted the internal NRC investigation, which concentrated on the mechanical and human causes. There were other studies and reports, most notably the report of the President's Commission on the Accident at Three Mile Island (Kemeny et al., 1979). President Jimmy Carter appointed the 12-person commission, chaired by John Kemeny, then president of Dartmouth College, because of great public concern about the TMI event. (See Walker, 2004, for a compelling description of the accident.) Budnitz says that his group focused on the technical and human factors, whereas the Kemeny committee focused on institutional and political causes of the event.

Three Mile Island: Causes and Lessons Learned

Robert Budnitz, Samuel Walker, and Buzz Savage each suggested descriptions of the accident. I have integrated and summarized their detailed notes to us in the following paragraph. In 1979, a complex sequence of events at the Three Mile Island Reactor Unit 2 (TMI-2) lead to a core damage accident. The incident began at 4 A.M. on March 28, 1979, when the coolant water to the pressurized water reactor was inadvertently cut off. Pressure in the reactor vessel quickly increased, causing the reactor to shutdown automatically, and a safety relief valve attached to the pressurizer to open. The reactor operators, overwhelmed by information from hundreds of display panels in the control room, did not understand fully what was happening. The automatic relief valve failed to close after the pressure was reduced, and water poured out through the opening. The emergency core cooling system (ECCS) responded automatically, as was intended, to keep the core covered with water. However, the operators, not recognizing that the relief valve was open, manually shut off the ECCS to prevent too much water from going into the core. Two hours elapsed before their mistake was discovered. Water continued to escape from the reactor, while the remaining water in the core started to boil. The water level dropped below the top of the core, causing the fuel elements to heat sufficiently to cause serious damage, releasing fission products into the primary system. The zirconium cladding that held the nuclear fuel pellets failed, and the fuel pellets collapsed into the bottom of the core and began to melt. Some of these radioactive materials escaped through the valve into the containment building, and some fission products were transported by coolant water into storage tanks in an auxiliary building. A few curies of radioactive gases were released through the containment buildings charcoal filters into the open air environment. The facility, however, including the primary system and containment, held the debris as it was supposed to. The TMI-2 events demonstrated the containment of the event extreme accidents and properly designed plants.

For the next 5 days, authorities in the state of Pennsylvania, the NRC, the White House, and other agencies sought to bring the reactor under control. Although it was apparent that a serious accident had occurred, no one knew how serious it was. The primary concern among policy makers and technical experts was that the containment building, which held the reactor, would be breached and that large amounts of the most dangerous forms of radiation would escape into the environment. The critical question was whether to order a mandatory evacuation of the population surrounding the plant. Governor Richard Thornburgh, who had to make the decision, was keenly aware that an evacuation could be costly in lives, injuries, and economic impact. But he also knew that if the plant failed, the result could be a public health disaster. Finally, after much anxiety and uncertainty, heightened by the discovery of a large hydrogen bubble in the reactor, it became clear that the accident would not cause a breach in the containment walls and spew volatile radioactive materials across the countryside of central Pennsylvania.

Budnitz attributes the immediate cause of the TMI event to equipment failure, compounded by misleading signals from the system, and operator error. As his group probed the TMI accident, they realized that the specific set of problems was symptomatic of more deeply seated human factor issues that needed to be addressed. Much of what Budnitz described to me were deeply rooted causes and lessons learned from exposing the roots and thinking about them. All are described in detail in Kemeny (1979) and the NRC reports and other assessments (see References at the end of this essay). Budnitz focused on what he considers the most important lessons learned. The primary lesson was "human factors" followed by equipment design and emergency planning. Budnitz summarized lessons learned in the following order. The TMI event was a wake-up call that "small" problems can lead to major consequences, in other words, every possible problem needs to be investigated. Prior to the TMI event, with regard to possible accidents, the NRC and nuclear utilities had focused their efforts largely on loss of coolant from a major pipe break. Well over $200 million (equivalent to about $1 billion today) was spent to understand this potentially catastrophic failure and ensure that the reactors can respond to it safely. The solution was to install and ensure the operation of automated backup coolant systems so that operator actions were not required to control a major pipe break. Yet, so much attention had been focused on large pipe breaks that, according to Budnitz, insufficient attention had been paid to what were perceived to be smaller problems.

The TMI event was one of these "smaller" problems. Budnitz observed that everyone thought a small pipe break and similar problems could be addressed by existing technology and operators. They were wrong. The entire industry quickly agreed, based on analysis done after the TMI accident that small problems were much more likely to happen than large ones. More important, the industry and the NRC both realized that every conceivable accident scenario needed to be identified and analyzed and then lessons learned developed to address each.

Another lesson learned was about operator training. Prior to TMI, operators had been trained on reactor simulators but not necessarily on a replicate of their exact reactor, because there were only a few simulators nationwide, whereas every reactor is different. Now, every reactor has its own simulator, which matches the on-site reactor and is kept up-to-date. The reactor simulators are located on-site, and one shift approximately every 2 weeks is devoted to working on the simulator.

A third lesson learned was about record keeping and the proactive use of write-ups of abnormal events to inform nuclear reactor owners and operators. Until the TMI accident, owners were required to report problems in the form of licensee event reports (LERs) within 60 days of the event. These LERs cover plant shutdowns, operations or other events that are not permitted, or an event that degrades the system, compromises safety, or both. Hundreds of LERs occurred every year, and the NRC staff studied them. However, the staff did not systematically look for patterns in the data, nor did they send lessons learned to sites. After TMI, the NRC established the Office for Analysis and Evaluation of Operational Data, which was charged with analyzing the event data and diffusing the information to the sites, where it can be used to improve safety, as well as efficiency. In addition, the Institute of Nuclear Power Operations (INPO) was created by the utilities in 1979. Headquartered in Atlanta, INPO rates the performance of every utility from 1 to 5. A rating of 5 is considered a sign of serious operational problems. These confidential ratings, according to Budnitz, have pressured companies into upgrading their systems and their operational capacity. More generally, INPO shares the industry's lessons learned, provides guidelines and best practices, offers workshops, and conducts on-site visits and training courses. Budnitz noted that the NRC, INPO, and economic pressures have led small utilities to sell their nuclear plants to larger utilities that have the expertise and capital to manage nuclear plants, which he feels clearly has increased safety, reduced risk, and increased plant availability and capacity factors.

Since TMI, the NRC has established many additional requirements and conducts more frequent inspections. Risk analyses are conducted before changes are permitted, to identify possible problems. Every reactor now has two to three dozen enclosed automatic signals that can detect problems and automatically shut down the reactor. Regulations exist to address a much larger set of plausible accidents. The NRC has established a presence at every power plant. Although some sites had resident inspectors, NRC staff from central locations would conduct many site visits. Now every site has one to three on-site inspectors who know the reactors intimately, and they are assigned to a different plant site very few years. At first, the utilities were opposed to on-site NRC inspectors but now they are fully supportive.

The NRC maintains a list of safety indicators for each facility. Two meetings, both of which are open to the public, are held every year at every site. Attendees can hear and read about the safety and efficiency of their local plant compared with that of every other plant. Anyone can compare the local plant with others and with the same plant at other times. Prior to TMI, NRC inspectors would conduct sporadic site visits. Now they are there all the time. Budnitz noted that public and business loss of confidence in the industry was justified after TMI, and that every one of the steps just described was essential to restore confidence, but especially this one. It was not a matter of choice; it was a matter of survival.

Looking back over the past 25 years, Budnitz believes that safety and efficiency have substantially improved. The worst performing plant today, he believes, performs better than the best performing plant at the time of TMI. He cited the number of "SCRAM" events or "trips" reported by the NRC as evidence. SCRAM events automatically shut down the reactor. Industrywide, he indicated that the number of automatic reactor shutdowns annually has dramatically decreased from about 1,000 a year at the time of TMI to about 50 or so. Equipment reliability has improved, operators are better trained, and surveillance by equipment, utility, and NRC staff is much improved.

Chernobyl: Causes and Lessons Learned

When the authors asked Budnitz, who visited the site, what could be learned from the Chernobyl disaster, he said little new could be learned about U.S. reactors because the Soviet reactors and operating policies were not comparable to those of the United States. The important lesson for the United States was the need for constant vigilance, a lesson that had been painfully learned at TMI.

Budnitz also described his response to Soviet management of their nuclear reactors at that time, comparing it to institutional control in the United States, where no reactor experiment can be performed without assessment of possible consequences and approval by two independent review committees. The Chernobyl reactor that discharged a substantial amount of radioactivity into the environment was an RBMK (reactor bolshoy mushchosty kanaly), a pressurized water reactor that uses water as its coolant and graphite as its moderator. The RBMK was unstable when operated at 15–25% power. But the operators did not know about (or ignored) this shortcoming, when they ran a test at 20% power that led to the large accident. They did not require approval to run that test. Samuel Walker told us that the operators were under pressure from their supervisors and party functionaries to conduct the test. Next, in the United States, a large power reactor would not be used for such a test. The Chernobyl operators, acting alone, used a massive power reactor for the test, and the operators turned off a primary safety system to determine whether a backup system would work; it did not. U.S. reactors have containments, typically 6 to 8 feet thick to contain a problem inside the facility. This Soviet reactor, which was originally designed to produce plutonium for nuclear weapons, not to produce electric power, had no containment. Also, Budnitz pointed out, U.S. reactors have multiple regulations governing operations of reactors; this Soviet reactor had very few, and many of those that they had were usually not honored.

Budnitz clearly is not a supporter of the old Soviet graphite reactor design. He explained his concerns by using an analogy to automobiles. One automobile is driving down a major thoroughfare in cruise control, and the driver periodically monitors the panels to make sure that they are functioning correctly. The other automobile must be monitored and adjusted every 5 seconds to make sure it stays on the highway. U.S. light water reactors are of the first, and the old Soviet graphite reactors are of the second group.

At the time of this writing, all four reactors at Chernobyl are permanently shut down. Lithuania also has shut down one of its two RBMK reactors and will shut down its second in 2009. The others, all in Russia, are near Smolensk (three), St. Petersburg (four), and Kursk (five) and are still operating. The Russians have made design changes, such as adding more manual control rods, adding absorbers to improve low-power operations, increasing fuel enrichment, and installing a separate reactor SCRAM system. Russian operators have been better trained to operate their graphite reactors. Budnitz remains concerned, however, because some of the institutional and engineering changes that were recommended have not been adopted by the Russians.

Buzz Savage added that there are no graphite moderated pressurized water reactors in the United States, and that all commercial power plants have the reactor system enclosed in a containment building. These decisions were made early in the development of nuclear power for peaceful applications, and were distinctly different than the Soviets development of nuclear power. He also recommended a report by the U.N. Scientific Committee on the Effects of Atomic Radiation (UNSCEAR, n.d.).

What Reporters Need to Know

Reporters have access to information about every U.S. commercial nuclear site and its on-site inspectors. They can attend biannual public meetings and ask questions. They may not be able to achieve a full understanding of the technologies but they should be able to derive information about the utility's diligence at managing the facility. Reporting this information to the public is important to the population, to the NRC, and to industry.

Pitfalls

TMI and Chernobyl were similar insofar as both were disasters for nuclear power. Yet the two were remarkably different. Chernobyl was a major disaster that revealed flaws in design, operation, and institutional control. It spread radioactive elements over a nation and a continent and caused other deaths and injuries because of the evacuations. The World Health Organization (WHO) has been gathering data and occasionally issues reports that can be accessed from its Web site (*www.who.int/en*). In comparison, from a public health perspective, TMI was not a "big" event, because emissions were minimal and epidemiological studies have so far shown no public health effects or ecological effects. (NRC studies show that the average dose was about 1 millirem compared with a natural background of 100–125 millirem a year for the area.) Americans and their European counterparts learned a lot from TMI and have dramatically reduced the chances of accidents. In Budnitz's opinion, the Chernobyl type reactors and management of them remain a major problem. In short, a generation after these accidents, he feels that the media should try to focus on the lessons learned in Western nations from TMI and much less on what happened at Chernobyl. Media should try to separate what happened as a result of the accidents.

References and Other Resources

Battist, L., Buchanan, J., Congel, F., Nelson, C., Nelson, M., Peterson, H., & Rosenstein, M. (1979). *Population dose and health impact of the accident at the Three Mile Island Nuclear Station: Preliminary estimates for the period March 28 through April, 1979.* U.S. Nuclear Regulatory Commission report. NUREG-0558. Washington, DC: U.S. Government Printing Office.

Bennett, B., Repacholi, M., & Carr, Z. (Eds.). (2006). *Health effects of the Chernobyl accident and special health care programmes.* New York: World Health Organization. *www.who.int/ionizing_radiation/chernobyl/who_chernobyl_report_2006.pdf.*

GPU Nuclear Corporation, Parsippany, NJ, home page. *www.firstenergycorp.com.*

Hore-Lacy, I. (2006). Light water graphite reactor (RBMK). In C. L. Cleveland (Ed.), *Encyclopedia of earth.* Washington, DC: Environmental Information Coalition, National Council for Science and the Environment. *www.eoearth.org/article/Light_water_graphite_reactor_(RBMK).*

Kemeny, J., Babbitt, B., Haggerty, P., Lewis, C., Marks, P., Marrett, C., et al. (1979). *Report of the President's Commission on the accident at Three Mile Island.* Washington, DC: U.S. Government Printing Office.

Three Mile Island Public Health Fund. Three Mile Island Citizens' Monitoring Network home page. *www.tmi-cmn.org/tmiphf.htm.*

United Nations Scientific Committee on the Effects of Atomic Radiation (UNSCEAR). (n.d.) *The Chernobyl accident: UNSCEAR's assessments of the radiation effects. www.unscear.org/unscear/en/chernobyl.html.*

U.S. Atomic Energy Commission. (1957, March). *Theoretical possibilities and consequences of major accidents in large nuclear power Plants.* WASH-740. Washington, DC: U.S. Government Printing Office.

U.S. Nuclear Regulatory Commission. (n.d.) 10 CFR. *50.73 licensee event report system. www.nrc.gov/reading-rm/doc-collections/cfr/part050/part050-0073.html.*

U.S. Nuclear Regulatory Commission. (1989, March). *The status of recommendations of the President's Commission on the accident at Three Mile Island Nuclear Station.* NUREG-1355. Washington, DC: U.S. Government Printing Office.

U.S. Nuclear Regulatory Commission. Division of Systems Research. Office of Nuclear Regulatory Research. (1990). *Severe accident risks: An assessment for five U.S. nuclear power plants.* NUREG-1150. Washington, DC: U.S. Government Printing Office.

U.S. Nuclear Regulatory Commission. (2004, March). *Fact sheet on the Three Mile Island accident. www.nrc.gov/reading-rm/doc-collections/fact-sheets/3mile-isle.html.*

U.S. Nuclear Regulatory Commission. (2007). *Fact sheet on the accident at the Chernobyl Nuclear Power Plant. www.nrc.gov/reading-rm/doc-collections/fact-sheets/fschernobyl.html.*

Walker, J. S. (2004). *Three Mile Island: A nuclear crisis in historical perspective.* Berkeley: University of California Press.

World Health Organization (WHO) home page. *www.who.int/en.*

Decommissioning Nuclear Facilities

Written by Michael R. Greenberg, based in part on an interview
with James H. Clarke, with comments by Eric Darois

Background

Before we review the decommissioning of nuclear facilities administered by the U.S. Nuclear Regulatory Commission (NRC), we note that there are differences between decommission of NRC-licensed nuclear power reactors (NRC licensed sites where radioactive materials were handled, including research reactors) and decommission of Department of Defense sites that were used during World War II for processing and production of nuclear materials. The decommissioning and remediation of these sites involves different waste forms, decommissioning approaches, and waste disposal solutions.

Every facility that has an NRC license to operate, including hospitals, nuclear power plants, low-level nuclear waste sites, and many others will eventually reach the end of its useful life. At that point, working with the licensee, the NRC will terminate its license and require that the facility be "decommissioned." Title 10 of the Code of Federal Regulations, section 50.2 (which applies to nuclear power plants) defines decommissioning as "the safe removal of a facility from service and reduction of residual radioactivity to a level that permits termination of the NRC license." The NRC's objective is to release the site for appropriate uses.

The NRC (2007b) divides sites into five categories: nuclear power reactors, research and test reactors, uranium recovery facilities, fuel cycle facilities, and complex sites. The last designation, complex site, includes locations where nuclear materials were used in various locations and ways at a single facility. The 2006 NRC report on decommissioning notes that approximately 200 materials licenses are terminated every year. Most of these, according to the NRC, are "routine." These license terminations are rarely for complex sites, and so no remediation is required.

In 2007, NRC decommissioning efforts focused on 16 nuclear power reactors, 14 research and test reactors, 32 complex decommissioning materials facilities, 3 fuel cycle, and 12 uranium recovery facilities. These facilities are in the process of being decommissioned or already are in long-term stewardship. The 16 power reactors include the well-known Three Mile Island Unit 2, and 15 others located primarily in the Northeast and on the West coast. The research and test reactors are at government facilities (e.g., NASA, Veterans Administration), universities (e.g., Cornell, University of Buffalo, and University of Illinois), and industrial sites (e.g., General Atomic, General Electric, Westinghouse). The complex decommissioning sites are located across the United States at military complexes, at companies, such as Westinghouse, Kaiser Aluminum, Englehard Minerals–Great Lakes, and many others, and the mines are owned by well-known companies, such as Exxon Mobil and United Nuclear, and less publicly recognizable ones, such as Bear Creek and Homestake. The 3 fuel cycle decommissioning facilities are owned by Areva NP, General Atomics, and Honeywell. A journalist interested in any of these sites or any of the more routine decommissioning activities can quickly find reports on the NRC Web site (*www.nrc.gov*).

At this point there is no permanent off-site destination for spent fuel. Hence, the spent fuel is left in pools on-site until it cools (usually about 5 years), and then it is moved on-site to dry storage in an independent spent fuel storage installation (ISFSI) where it is stored under a new NRC license or under the existing operating plant license. The physical plant is typically decontaminated (DECON) and then torn down, although other options, such as entombment (ENTOMB) or safe storage (SAFESTOR) are available.

A good way of envisioning what happens is to view decommissioning as the slow motion reverse of constructing a new building. That is, the architects and engineers, instead of designing and then building the facility, start with the existing building and grounds and design a process to take the buildings down and remediate the contamination. Decommissioning a facility of the size of a nuclear power plant takes a long time, typically 5 to 10 years, and the costs have averaged over $400 million (reported costs inflated by the Consumer Price Index to 2007 dollars by the author). (See NRC, n.d., 2007a, 2007b; WNO, 2007, for non-U.S. experience.)

NRC requires that the licensee have funds set aside for decommissioning. One method is prepayment, for which the owner deposits funds in an account that remains outside its cash or liquid assets. In this case, a utility cannot change its mind and take the money back. A second method requires periodically setting aside funds into an account, and these accumulated funds must be sufficient to cover anticipated decommissioning costs. Third, the licensee can post a bond or letter of credit (which must be renewed). Information about the various methods available to the licensee is available in NRC documents.

The decommissioning process is carefully scripted and under the oversight of the NRC, although 34 states have formal agreements with the NRC, permitting the states to exercise regulatory control over some quantities of nuclear materials and decommissioning of some complex material sites. These agreements do not extend to the decommissioning of nuclear power plants, which are the responsibility of the NRC.

Decommissioning of a nuclear power plant begins when a licensee chooses to cease operations at a site. The plant owner is required to submit a written notification to the NRC within 30 days of its decision to cease operations. It must notify the NRC when nuclear fuel has been removed from the reactor as part of decommissioning. The next stage is to submit a post-shutdown decommissioning activities report (PSDAR). The PSDAR must be submitted within 2 years of ending site operations, and it must include a description of the planned decommissioning activities, a schedule for these activities, an estimate of expected cost, and a review of environmental impacts associated with decommissioning. The NRC is required to publish the PSDAR in the *Federal Register* and to hold public meetings near the site.

Next, the applicant must submit a license termination plan (LTP), which describes the site in detail, dismantling of the facilities, plans for site remediation, plans for a radiation survey, the end use of the site, and updates on estimated decommissioning costs and environmental impacts. Once the NRC receives the LTP, it is made available for public comment. And once the LTP is approved, it will be implemented, and NRC staff will monitor the decommissioning process. At the end of the decommissioning process, the site owner must submit a final status report that reviews what has been accomplished. This process is both tedious and expensive, and federal government rules allow decommissioning to take place over 60 years.

The record of the 16 power reactors undergoing decommissioning illustrates just how long the process takes. The Humboldt Bay site formally entered SAFESTOR in the 1980s. A PSDAR was submitted for Three Mile Island (TMI) Unit 2 in 1979. However, the owners have applied for a 20-year license extension for Unit 1 and current plans defer final decommissioning of Unit 2 until decommissioning of Unit 1 is complete. PSDARs for Big Rock Point in Michigan and Connecticut Yankee were submitted in 1997. Their decommissioning was completed in 2007.

Identifying the Issues

One concern is how clean is clean. The NRC's requirement for license termination to unrestricted release is that a site cannot be abandoned unless the residual radioactivity results in an annual dose of 25 millirem (mrem) or less. Utilities are allowed to apply for "restricted" release (restricted use), which means that it must use fences, federal government ownership, or other mechanisms to prevent public access. The owner can also argue for a 100 mrem or 500 mrem standard, but then institutional control requirements become a much bigger issue. In these latter cases, the utility must persuade the NRC that environmental or human impacts will result from attempting to meet the 25 mrem standard, that they have consulted with surrounding communities, and that they have placed sufficient funds in accounts outside their control to pay for institutional control. In the case of 500 mrem, they must demonstrate, in addition, that the lower achievable level is not economically feasible. For context, background radiation in the United States typically is 250 to 350 mrem a year. Note, however, that no commercial nuclear power plant has applied for a restricted release or for higher release levels. Some local officials and residents are likely to assert that no residual contamination be left. Related to both radiological state and status of spent fuel is future site use after decommissioning. If the site is classified as unrestricted, presumably any use is permitted, including farming, natural habitat, and industrial activity. If the site is left as "restricted" use, then NRC approval is granted case by case.

From the owner's perspective, to eliminate all of the residual is an enormous expense for little risk reduction. In short, the final radiological state of the site will be an issue that will interest local officials and communities, but so far the industry has not requested a change in the NRC's standards. It will be an issue to follow, but it may not turn out to be a real story.

A second issue is the destination of the spent fuel. As described elsewhere in this volume, the original intent was to move the spent fuel to a permanent repository at Yucca Mountain. This proposal has become a contentious issue in the United States that does not appear to be resolvable in the next 5 years, if at all. Another option is to remove elements of spent fuel and reuse them in nuclear power plants. This option is incorporated into the Global Nuclear Energy Partnership (GNEP), which, like Yucca Mountain, is controversial. (See "Closing the Civilian Nuclear Fuel Cycle.")

What Reporters Need to Know

From the declaration of intent to close a site through the final stages of the process, reporters have access to the process and the documents. NRC licensing actions require public hearings. A reporter can keep track of this lengthy and meticulous process by looking at the NRC's annual report on the decommissioning status of every plant, which is a one- or two-page summary of what has gone on and what is planned. Reporters should be able to determine how much, if any, radioactive material has been left, and how it is being safeguarded. The final survey documents are difficult to read, requiring considerable technical preparation, but reporters will be able to gain access and understand conclusions that the NRC makes by reviewing the final survey reports and asking questions.

More specifically, U.S. nuclear power plant owners must make a case for three decommissioning approaches. One is decontaminating and dismantling parts of the reactor soon after the plant closes. A second approach is to let the plant remain for up to 60 years, which allows radiation to decay. In this case, the facility is monitored until radiation is decreased to lower levels, then it is dismantled. The third option is to entomb the radioactive elements in concrete and steel for a longer time, which isolates the sources of radiation and radioactivity from the public and the environment. Licensees can pick one or a combination of these three options for decommissioning. The licensee's strategies and justifications should be reflective of the particular conditions and constraints that a utility faces at the time of these decisions. The decisions might make an interesting story about the history of a local plant and its management and future use.

It is fair to say that the NRC has been among the most scrutinized organizations in the U.S. government. In addition to checking on safety issues associated with the local plants, some journalists may be interested in NRC's ongoing efforts to improve its performance. Unlike many federal organizations, this one, because of its past history and its role, has an extensive lessons-learned program that it shares. For example, its "decommissioning lessons learned" report covering September 16, 2004, through October 20, 2006, briefly describes 23 lessons learned and actions taken. Some of these are on data management and other issues that are not necessarily newsworthy. But others reveal NRC's efforts to be more transparent in dealing with licensees and other interested parties about how it wants to interact during the decommissioning process (NRC, n.d.).

Pitfalls

Two issues surface during decommissioning. The first is the NRC's 25 mrem per year criteria requirement. That number must be put in the context of background radiation exposure in news stories to ensure that readers understand its significance. Second, the issue of leftover spent fuel sometimes surfaces, as it should. However, the process for on-site storage of spent fuel too often is portrayed as an issue for the Yucca Mountain area of Nevada. In fact, there are immediate safety, security, and other local concerns resulting from the reality that spent fuel is stored on-site. Until the fuel and spent fuel is removed, little can be done about planning the future of the location on that part of site where the spent fuel is stored, although other parts of the site can be reused.

References and Other Resources

National Energy Institute. (2007). *Decommissioning of nuclear power plants. www.nei.org/keyissues/nuclearwastedisposal/factsheets/decommissioningnuclearpowerplants.*

Organization for Economic Co-operation and Development. Nuclear Energy Agency. (2004). *Strategy selection for the decommissioning of nuclear facilities: Seminar proceedings, Tarragona, Spain, 1–4 September 2003.* Allan Duncan. Paris: OECD Publishing.

U.S. Nuclear Regulatory Commission (NRC) home page. *www.nrc.gov.*

U.S. Nuclear Regulatory Commission. (n.d.) *Decommissioning lessons learned. www.nrc.gov/about-nrc/regulatory/decommissioning/lessons-learned.html.*

U.S. Nuclear Regulatory Commission. (2007a). *Fact sheet on decommissioning nuclear power plants. www.nrc.gov/reading-rm/doc-collections/fact-sheets/decommissioning.html.*

U.S. Nuclear Regulatory Commission. (2007b). *Frequently asked questions about reactor decommissioning. www.nrc.gov/about-nrc/regulatory/decommissioning/faq.html.*

U.S. Nuclear Regulatory Commission. Office of Federal and State Materials and Environmental Management Programs. (2007c). *Status of decommissioning program, 2006 annual report.* NUREG-1814-, Rev.1. Washington, DC: U.S. Government Printing Office.

World Nuclear Organization (WNO). (2007). Decommissioning nuclear facilities. *www.world-nuclear.org/info/inf19.html.*

Transportation of Nuclear Waste

Written by Bernadette M. West, based in part on an interview with
Mark Abkowitz, with comments by Penelope A. Fenner-Crisp

Background

Transport is a critical part of the discussion involving nuclear materials and the waste generated in the course of their use. Most nuclear materials are transported several times during their lifetime. For example, nuclear materials used in the fuel cycle are transported from mining to milling, from milling to conversion to enrichment, from enrichment to fuel fabrication to power generation, with the spent fuel then moved to temporary storage on-site and eventually to permanent storage off-site.

In the United States today, spent nuclear fuel and high-level radioactive waste is temporarily being stored at 131 sites in 39 states, including U.S. Department of Energy (DOE) environmental cleanup sites as well as commercial facilities that are continuing to produce spent fuel while generating the nation's electricity supply. Waste generated by these facilities must be transported in order to proceed with environmental cleanup at DOE sites and to create space for new waste continuously being generated at commercial sites.

In the United States nuclear waste is transported every year, primarily by highway, rail, and water (although sometimes by air), to temporary and permanent repositories, depending on the type of waste. The volume of these shipments could notably increase in the future if the industry were to expand a great deal and if a single permanent repository for nuclear waste were to be opened.

The transportation of nuclear waste is regulated jointly by the U.S. Nuclear Regulatory Commission (NRC) and the U.S. Department of Transportation (DOT). The NRC sets standards for packaging design and performance standards as well as protecting shipments while in transit. They regulate the testing of a shipment before it leaves a site to check for leaks and to ensure that radiation levels are within safe limits. The NRC works together with 32 states that have agreements with the NRC for handling certain nuclear waste.

While nuclear materials are in transport, the DOT then handles regulation of the shipment. They are responsible for training personnel who handle the transport (including drivers), as well as for labeling, shipping papers, placarding, loading, and unloading. DOT establishes control route selection and is responsible for inspecting vehicle condition.

The packaging, storage, and transportation of radioactive waste from nuclear weapons production and site cleanup by the DOE follows standards set by the DOT and the NRC.

For more detailed information, including some of the history about the requirements and processes, our experts recommend examining the Web sites of the Licensing Support Network (*www.lsnnet.gov*), which supports the NRC's missions, and the Nuclear Waste Technical Review Board (*www.nwtrb.gov*), which is an agency of the federal government charged with independent scientific and technical oversight of the DOE's program for managing and disposing of high-level radioactive waste and spent nuclear fuel.

Identifying the Issues

Transportation of nuclear waste must be understood in terms of the different types of radioactivity that have been and still are being generated. Spent fuel is produced when electricity is generated in nuclear power production. Nuclear fuel generates heat when it undergoes fission inside a nuclear reactor and this heat is used to generate steam, which then passes through a turbine and turns a generator. When nuclear fuel no longer generates enough heat to generate electricity it is considered "spent," although as noted several other places in the book, it can be reused. The highly radioactive spent fuel remains contain fission products, unused uranium and plutonium, and various other transuranic (TRU) elements with an atomic number greater than that of uranium. Approximately 1% to 5% of the fissionable uranium in a light water reactor is used before the reactor is refueled with fresh uranium. The waste must be removed from the reactor and put into spent fuel cooling pools inside the containment building until the spent fuel is safe to handle. It is then certified by the NRC for interim on-site storage, and, finally, it is prepared for transport in an NRC certified transportation cask and placed in a repository for safe underground disposal. Spent fuel comes from commercial nuclear power plants, domestic research reactors, nuclear-powered U.S. naval warships, DOE-run research and defense reactors, reactor design testing, and energy and medical research.

High-level waste (also referred to as HLW) results from the reprocessing of spent nuclear fuel. It includes liquid waste produced directly during reprocessing of spent fuel to recover usable uranium and plutonium. High-level waste is highly radioactive and can be thermally hot. It contains fission products, traces of uranium and plutonium, and other TRU elements. It also includes solid material derived from liquid wastes that contain fission products in sufficient concentrations. Other radioactive materials determined by the NRC to require permanent isolation are also considered high-level waste. The reprocessing of spent fuel from reactors ended during the Carter administration. High-level radioactive waste must be isolated for thousands of years during its decay process until it reaches safe levels.

DOE has responsibility for the transportation of spent fuel from commercial nuclear power plants and Department of the Navy and DOE reactor sites, as well as high-level radioactive waste from DOE and defense-related sites, to a permanent geologic repository for disposal, when and if a permanent site becomes operational. Finding a permanent storage site for high-level waste is still under discussion, with Yucca Mountain, Nevada, having been identified as the exclusive site under consideration.

At present, commercial spent fuel is being stored on site at power plants in pools of water or in concrete or steel casks, otherwise known as dry storage. These are located outside the containment building. Spent fuel and high-level radioactive waste from defense programs are being stored in underground tanks and in dry storage at DOE and other government sites.

Water is often removed from high-level waste and it is then solidified. The high-level waste is made into solid small ceramic pellets (approximately the size of a pencil eraser) and enclosed in strong, multi-layer metal tubes that are specially made to contain radioactive materials during use inside a reactor and also for long-term storage. High-level waste from defense programs is sometimes solidified into glass, which makes it highly resistant to water. In some cases, for example, at the Hanford K-basins in Washington, the solid waste is de-watered and mixed with concrete for disposal; in other cases, for example, at West Valley, New York, the solid waste is mixed with silica and vitrified into pellets or logs. The ultimate goal is to prevent the high-level waste from being leached into groundwater. De-watering reduces both the volume of waste and the risk, because solid waste is much less mobile than liquid and gaseous waste.

Low-level waste (also referred to as LLW) is generated by industries, hospitals, universities, the fuel-making process, and research facilities. It includes used protective clothing items, rags, mops, filters, reactor water treatment residues, equipment and tools, luminous dials, medical tubes, swabs, injection needles, syringes, and laboratory animal carcasses and tissues. The amount of radioactivity in low-level waste varies—from trace amounts that are at or near levels found in nature to much higher levels in water treatment residues, small gauges containing radioactive materials, and materials taken from inside reactor vessels in nuclear power plants. Higher levels require some shielding and transport. Very low-level waste can be buried on site until the radioactivity has decayed sufficiently. Then it can be disposed of like ordinary trash, although it must comply with stringent state and federal monitoring and surveillance regulations. Higher level low-level waste is allowed to accumulate and then is shipped in special containers to low-level waste disposal sites. The DOT and NRC regulate transport of low-level waste. Currently low-level waste can be transported to three commercial land disposal sites—Richland, Washington; Clive, Utah; and Barnwell, South Carolina. These sites accept only certain types of waste and from only certain states, and it is unclear how long they will continue to accept waste. Barnwell, for example, has indicated that it will stop out-of-state shipments, with a few exceptions. Everything else is stored on site where it has been produced, such as at hospitals, research facilities, and clinics and nuclear power plants. Shipping of low-level waste must be done in containers approved by DOT. Previously, low-level radioactive waste was shipped to four other sites—near Sheffield, Illinois; Morehead, Kentucky; Beatty, Nevada; and West Valley, New York. These four sites have now been closed and no longer accept waste.

Transuranic waste (TRUW) is generally produced during weapons production and includes contamination with alpha-emitting radionuclides. It has a very long half-life. The United States currently permanently disposes of TRUW at the Waste Isolation Pilot Plant near Carlsbad, New Mexico. (See "Closing the Civilian Nuclear Fuel Cycle" for a discussion of radiotoxicity.)

Mill tailings waste results when uranium and thorium are extracted from natural ore. This waste—sometimes called 11e.(2) waste—can be found at mining sites in Colorado, New Mexico, Utah, and elsewhere. The waste is usually stored at or near where it was produced. It is buried and covered with clay, soil, rocks, or other materials to limit erosion and prevent radon from escaping. These actions reduce the release of radon but do not entirely prevent it. Mill tailings are transported to the Clive, Utah, site for temporary storage.

While federal agencies regulate the transportation of nuclear waste, it is important to remember that transportation of such waste does not occur in a governmental or social vacuum. States and local and tribal governments must also play a role, cooperating with federal agencies in transportation decisions, including the identification of preferred routes.

Containers for Transporting Waste

Containers for packaging radioactive materials must meet rigid NRC design specifications and testing requirements. These casks are designed to shield radiation and contain radioactive materials. The NRC must approve all packages used for shipment. It issues a radioactive material package certificate of compliance (CoC) to a company if the package meets all requirements.

Nuclear materials are shipped in several types of containers, with the type dependent on the level of radioactivity involved and the form of the material to be transported. Low-level waste may be shipped in any of three different types of containers, all of which must meet both NRC and DOT requirements. Very low-level radioactive materials can be shipped in strong, tight containers. Materials that provide little hazard from radiation exposure, such as smoke detectors, can be shipped in industrial packages. These containers used for shipping such materials as medical or industrial radioisotopes have been determined to withstand normal transportation activities. Higher level low-level waste must be shipped in either type A or type B containers. Most low-level waste is shipped in type A containers. They are heavy steel drums or boxes, tested to be sure they can withstand normal transportation conditions. Type B containers, which are heavy metal engineered casks, are used to transport materials with higher levels of radiation. They are tested under both normal and accident conditions. This includes demonstrating an ability to survive a 30-foot fall, a 40-inch drop onto a 6-inch spike, exposure to fire for 30 minutes, and submersion in 50 feet of water for 8 hours.

Spent fuel and high-level waste are transported in type-B transportation casks that are then put into overpacks. The design of these casks is regulated by the NRC, as described in Title 10, Part 71, of the Code of Federal Regulations. Casks are typically made from stainless steel (usually 3 inches thick) with metal shielding that is usually more than 6 inches thick. They are designed to contain radiation during routine transporting as well as to withstand impact, puncture, fire, and submersion under water. Casks are much thicker than a gasoline tank truck shell. Truck containers weigh approximately 25 tons when loaded with 1–2 tons of spent fuel. Rail containers are heavier, weighing up to 150 tons, and can carry up to 20 tons of spent fuel. Impact limiters are placed on the ends of the containers, making them look like concrete dumbbells. Impact limiters are designed to absorb the force of impact in the event of an accident. While casks are tightly sealed, it is not possible to eliminate all radiation with shielding. However, the "normal" amount of radiation is reduced to low levels that meet standards set by DOT and NRC. Container designs must be tested using a computer analysis as well as actual physical testing prior to the issuance of a CoC to the manufacturer.

DOT requires trucks carrying nuclear waste (class 7 radioactive material) to be labeled as such. Labels range from I to III, with larger numbers corresponding to higher levels of radiation at the surface of the package and therefore having the need for greater precaution to ensure safe transport. Labels are placed on the rear of a truck. The lowest category is RADIOACTIVE WHITE-I and the highest is RADIOACTIVE YELLOW-III. For example, a package with a maximum surface radiation level of 0.6 millisievert (60 millirems) per hour must bear a RADIOACTIVE YELLOW-III label. Above certain shipment volumes, packages must be marked as "highway route controlled quantity" and must be labeled RADIOACTIVE YELLOW-III.

DOT regulates transport of TRUW in type B containers. Over 90% of TRUW being transported to the waste isolation pilot plant (WIPP) is categorized as "contact-handled." This means that it emits primarily alpha and beta radiation that can be handled under controlled conditions without special shielding. Contact-handled waste can be transported in 55-gallon drums and boxes. Stainless steel containers are used to transport "contact-handled" materials by truck to the WIPP site. These containers are constructed with inner and outer containment vessels. Remote-handled waste emits gamma radiation and must be transported in specially designed lead- and steel-shielded containers with another thermal shield called RH-72B. These containers are then placed inside circular impact limiters that are similar to shock absorbers. For waste containing higher concentrations of fissionable materials, such as plutonium-239 and uranium-235, a special "pipe overpack" container is used to separate and restrict the quantity of fissionable materials that are transported so that a critical mass cannot be formed under any postulated accident conditions. This overpack is first placed within an impact limiter before it is enclosed in a 55-gallon drum and placed in the containment vessel.

Modes of Transportation

Truck, train, and barge are the three basic modes used to transport nuclear materials. In addition, a limited number of small volume shipments are transported by air. Almost all low-level waste is moved by truck. High-level waste and spent nuclear fuel are often moved by rail and sometimes by truck where rail is not available. Often a shipment involves a combination of modes; these are referred to as intermodal shipments. Much of the current transport that occurs is between different reactors owned by the same company who are sharing storage space.

Although truck transport is more common, rail shipments can accommodate the movement of larger loads. For shipments of spent fuel and high-level waste to the proposed Yucca Mountain site, DOE has proposed shipping primarily by rail using dedicated trains—where the shipment of nuclear waste would be the only commodity on the train. This proposal is intended to prevent the possibility that a spent fuel or high-level radioactive waste release could interact with other hazardous commodities, such as flammable liquids, on the same train. Trucks and barges would be used to connect waste generation sites to a rail access point in places where rail transportation does not exist.

Routing

Routing guidelines for transporting nuclear materials are under the jurisdiction of the DOT. Criteria for selecting routes vary according to the type of shipment. Spent fuel and high-level waste are classified as "highway route controlled quantity." DOT requires carriers of waste with this classification to use "preferred highway routes," which aim to minimize transit time and avoid population centers to the greatest extent possible, thereby limiting exposure to the public. Routes generally use the interstate system and, where available, use bypass routes that go around cities. In some cases, alternative routes have been officially designated by the states. A variety of factors are considered when establishing highway routes, including accident rate, transit time, time of day, population density, and number of other vehicles sharing the road. Limiting use of tunnels is another consideration. A global satellite communications network keeps trucks and dispatchers connected at all times. Trucking companies that handle many different kinds of hazardous materials are used to transport the nuclear materials. Drivers receive special training and become certified for their role. They must carry copies of their certification papers while en route. Our experts believe that armed escorts are planned, although final requirements have not been set.

Shipments by rail differ from highway transit in that historically rail lines were built to connect population centers, so they usually pass through urban centers. Because railroads are also privately owned, rail carriers have greater discretion over how to route a particular shipment.

Shipments must use NRC-approved routes, have procedures in place to address emergencies, maintain log books, remain in contact with a communication center, and notify local law enforcement officials en route. The governor of each state through which shipments pass must be notified in advance. All shipments must have 24-hour satellite tracking and must be accompanied by escorts who report to a central command facility every 2 hours. The NRC must coordinate movement of shipments with state or tribal officials, law enforcement and emergency response officials, and others on a need to know basis. For security reasons, however, the specific timing and routes of shipments is kept classified from others. The NRC is required to document oversight of shipments in an inspection report, which is available to the public on the agency's document management system (ADAMS). Violators of the shipping regulations are issued sanctions that include notice of violation, fine, requirement for modification, suspension, and revocation of a license.

NRC regulations for shipments carrying nuclear waste to the proposed Yucca Mountain site will require carriers to follow only approved routes. They will also be required to provide armed escorts for heavily populated areas and coordinate with local law enforcement before shipping is under way. States through which shipments will pass must be notified in advance of the shipment. Casks must be secured and carriers must provide monitoring and communications at all times. Satellite tracking will be used.

Should the proposed Yucca Mountain site become operational, the DOE estimates there will be an average of 130 rail and 45 truck shipments annually for 24 consecutive years. That would be the equivalent of about 1 shipment every 2 days somewhere in the United States.

There are approximately 300 million shipments of hazardous materials annually in the United States. Of these, 3 million involve radioactive materials shipped to hospitals, universities, industrial and manufacturing plants, research facilities, and radioactive waste disposal facilities. The rest involve explosive or highly flammable and other toxic materials. The proposed spent nuclear fuel shipments to Yucca Mountain (175 per year for 24 years) would represent less than 0.01% of all radioactive shipments currently on routes crossing the United States today. Approximately 95% of this would be shipped by rail—not highway or barge.

Training and Emergency Response Planning

States and tribes receive federal support for special training for local officials along transportation routes in preparation for nuclear shipments. Emergency responders are trained by DOT, DOE, and the Federal Emergency Management Agency (FEMA) for a wide variety of potential incidents and accidents. WIPP personnel have been training state and tribal first responders and emergency medical personnel in states since 1988.

FEMA has developed the Federal Radiological Emergency Response Plan. The plan coordinates the efforts of all government agencies and involves local police, firefighters, and state radiological protection teams who would be involved in responding to a transportation accident. Nuclear utilities have agreed to provide backup support should an accident occur involving transportation of radioactive materials.

In addition to the Federal Radiological Emergency Response Plan, each transporter must also have in place an emergency management plan that lays out response actions to be taken in the event of an accident. The plan lists the roles and responsibilities of local, state, and federal agencies that would respond to accidents. It includes emergency contact information for involved parties, including contacts in all states along the route. In addition to the emergency management plan, shippers need to have a carrier transportation plan that details carrier resources available to assist in response and recovery plans. Occupational Safety and Health Administration (OSHA) regulations require that first responders be trained for emergencies involving radioactive materials. First responders must be able to recognize that an accident involves radioactive materials, isolate the scene, and call for assistance from specially trained personnel.

What Reporters Need to Know

When considering conditions for safe transport of nuclear waste, one needs to differentiate between three types of risk:

- "Incident-free" risk involves exposure to the public that occurs from shipments under normal (nonaccident) conditions. Despite all the safeguards put in place in the design of containers and overpacks, there is still a small amount (at safe levels) of radiation that emanates from the package. Any potential risk here would more likely be to workers who handle the actual materials than to the general public on the routes through which shipments are passing.
- "Accident without release" risk refers to the potential for a vehicle (truck or train) accident to occur where the accident does not involve release of the commodity. This risk is the same for any freight movement on a particular mode, irrespective of the type of freight being moved, because the commodity itself does not cause any harm. Rather, damage consists of impacts to human health, property, and the environment from the collision itself.
- "Accident with release" risk refers to an accident that involves a release of the contents of the container. Here, the potential harmful effects of the commodity itself are also taken into consideration, since the environment in proximity to the accident site is directly exposed to the material. For this reason, this type of risk poses the greatest concern.

In a situation involving a transportation shipment of nuclear waste, it is critical to differentiate between accidents that involve no release and those that involve a release. The risks to the public are very different for each of these scenarios. Where no release occurs, an accident involving a shipment of nuclear waste does not automatically translate into an exposure for the community. Even when a release does occur, it is important to ask about the size of the container breach and the release rate. A small hole and a low release rate could pose little threat, whereas a large hole and a high release rate could have severe consequences.

There are several properties of spent nuclear fuel that are important when considering the issue of transportation:

- Spent fuel is not a liquid or a gas, and it will not pour or evaporate. However, it contains uranium dioxide (Buzz Savage, personal communication). At high temperatures (e.g., above 300°C), uranium dioxide will undergo additional oxidation when exposed to air into a more stable oxide. In the process, the uranium solid disintegrates into a powdery form and may be dispersed into the environment unless contained within the transportation cask.
- Spent nuclear fuel does not burn—even if it is engulfed in flames. It is not flammable.
- Spent nuclear fuel cannot explode. It is not explosive and is not a "bomb" of any sort—nuclear or otherwise.

Pitfalls

While a primary concern for safety in transporting nuclear materials is in the design of packages that can withstand accidents, this aspect represents only part of the equation. It is also critical to plan and manage shipments with stringent regulations that are carefully enforced over the entire period of the shipment.

When DOE is the shipper, state and local governments still have primary responsibility to protect the health and safety of the public.

Mill tailings—waste from mining uranium—may contain other hazardous chemicals, as well as the heavy metals lead and arsenic.

Resources

Council of State Governments Midwestern Office. (2005, August). *Handbook of radioactive waste transportation. www.csgmidwest.org/About/MRMTP/PublicInformation/realHandbook_inside.pdf.*

Licensing Support Network home page. *www.lsnnet.gov.*

National Research Council. Committee on Transport of Radioactive Waste. (2006).

Going the distance? The safe transport of nuclear fuel and high-level radioactive waste in the United States. Washington, DC: National Academies Press. *www.nap.edu/catalog.php?record_id=11538.*

U.S. Department of Energy. *Transportation Emergency Preparedness Program (TEPP). www.em.doe.gov/Transportation/TEPP_Home.aspx.*

U.S. Department of Energy. Waste transportation, general information. *www.energy.gov/safetyhealth/wastetransportation.htm.*

U.S. Department of Energy. Office of Civilian Radioactive Waste Management. *Transportation of spent nuclear fuel fact sheet. www.ocrwm.doe.gov/factsheets/doeymp0500.shtml.*

U.S. Department of Energy. Office of Public Affairs. *Spent nuclear fuel transportation. www.ocrwm.doe.gov/transport/pdf/snf_trans.pdf.*

U.S. Department of Transportation. Federal Motor Carrier Safety Administration. *Part 397: Transportation of hazardous materials; Driving and parking rules. www.fmcsa.dot.gov/rules-regulations/administration/fmcsr/fmcsrguidedetails.asp?rule_toc=766§ion_toc=766.*

U.S. Department of Transportation. Research and Special Programs Administration. *Federal regulations on labeling hazardous materials for transportation. ecfr.gpoaccess.gov/cgi/t/text/text-idx?sid=585c275ee19254ba07625d8c92fe925f&c=ecfr&tpl=/ecfrbrowse/Title49/49cfrv2_02.tpl.*

U.S. Nuclear Regulatory Commission. Nuclear materials transportation, general information. *www.nrc.gov/materials/transportation.html.*

U.S. Nuclear Regulatory Commission. *Nuclear waste transportation routes: Highway and rail routes most likely to be used to transport high-level nuclear waste to Yucca Mountain, Nevada. www.state.nv.us/nucwaste/states/us.htm.*

U.S. Nuclear Regulatory Commission. Spent nuclear fuel transportation, general information. *www.nrc.gov/waste/spent-fuel-transp.html.*

U.S. Nuclear Waste Technical Review Board home page. *www.nwtrb.gov.*

The Economics of Nuclear Power

Written by Bernadette M. West, based in part on interviews with
Paul L. Joskow and Henry J. Mayer, with comments by Seth Blumsack

Background

There are currently 104 commercial nuclear power plants in operation in the United States and, as of August 2007, there were 439 worldwide. In the United States, 7 reactors are in the planning stage and 25 new plants have been proposed. No new plants have received construction permits in the United States since 1979.

Higher fossil fuel prices, improvements in plant design that are expected to lower construction costs for new plants and make them more predictable, revisions in the federal licensing process that have reduced the time needed for licensing, new federal subsidies as a result of the federal 2005 Energy Act, and the need to reduce carbon emissions—are all contributing to an environment that seems more encouraging to new nuclear power plant construction. For detailed reviews, we recommend Mazuzan and Walker (1985) and Walker (2000).

Meanwhile the licenses of existing nuclear power plants are set to expire in the next 30 years. Since nuclear plants have high utilization factors, they represent approximately 20% of both energy (MW-hours) and capacity (MW) in the United States.

If they are not extended or renewed, the percentage of electricity generated by nuclear power will drop dramatically over the next 30 years. Many existing plants have applied for and received extensions that would allow them to operate for an additional 20 years beyond their initial 40-year licenses. Capital costs to extend the lifetime of existing nuclear plants is considered to be much less than construction costs for new coal or natural gas combined cycle gas turbine (CCGT) plants. Upgrades to existing plants to produce more electricity—while possible—will likely only yield small additional amounts—3% at best.

As industrialized countries wrestle with how to curb carbon dioxide emissions, nuclear energy has one indisputable advantage over coal, oil, natural gas, and even biological fuels: it emits no carbon dioxide. Global pressure to reduce CO_2 emissions and more aggressively address global warming concerns has led to increased interest in the United States in promoting investment in new nuclear power, but there is continued economic uncertainty. In the future, society faces a number of choices in meeting its electricity needs, including moving forward with building new nuclear plants; extending the life of existing nuclear plants and giving them power upgrades to produce more; relying more on coal and gas to generate electricity, processes that produce greenhouse gases and contribute to global warming; or switching to alternative fuels such as wind.

Identifying the Issues

While natural gas and coal prices were falling in the late 1980s and 1990s, nuclear power plants faced numerous problems that drove up their economic costs. These included lengthy periods for construction and construction cost overruns, long licensing processes, high operational costs, community opposition, and uncertainties regarding how to handle spent fuel and permanent storage facilities.

These issues were less problematic in the past when plants were built primarily by regulated investor-owned vertically-integrated utility monopolies, as in the United States, Germany, Spain, and Japan, or by state-owned electric power monopolies, as in France, the United Kingdom, Canada, India, and the former Soviet Union. In the past, consumers—not providers—were made responsible for handling the risks of high construction costs, lengthy licensing processes, shifts in fuel prices, and other uncertainties. Investors were protected and monopolistic providers could guarantee output requirements.

Today, in Europe and portions of the United States and Canada, the electricity sectors are being restructured to rely more on competitive power markets. While traditionally regulated utilities, such as Southern and TVA, are considering new nuclear units (or are likely to consider new nukes), many future nuclear power plants will be built by private firms subject to conventional market risks regarding costs, operating performance, and power prices. In such a competitive marketplace, investors will bear the risk of uncertainties associated with getting permits and with construction costs and operating performance, and private generating companies will face lower-cost competition. New generation plants will need to compete on the basis of cost, without large government subsidies. Those with lower capital costs—despite higher fuel prices—will be able to provide quicker returns on investment. In some places, new plants will compete with alternative fuel plants in a deregulated market; in others, new plants will still operate in a regulated environment but under conditions that require competitive bidding and greater risk sharing with investors.

The possibility of building new nuclear power plants raises a number of questions (some as yet unresolved) about the economic viability of nuclear power in this new environment. Certain challenges to the future of nuclear power have been reduced, such as an attempt by the U.S. Nuclear Regulatory Commission to reduce uncertainties in the licensing process, and benefits, such as loan guarantees to assist the owner in obtaining favorable interest rates on borrowed money, and insurance to reimburse the owner for unanticipated delays in the licensing process. These benefits are specified in the Energy Policy Act of 2005. The act, which has helped create an economic environment more favorable to new construction of nuclear power plants, provides for loan guarantees for up to 80% of project costs and a production tax credit of $18 per Mw (megawatts electric) for new nuclear capacity and insurance protection on construction delays for up to 6,000 Mw of new "first of a kind" nuclear generating capacity. Government loan guarantees do not reduce construction costs, but they do appreciably reduce the risks associated with the construction of new nuclear plants. Particularly in the absence of a price on carbon emissions, these loan guarantees may tip the scales in favor of some proposed nuclear projects.

A large part of the reason for high capital costs in the previous generation of nuclear power plants was that each plant was essentially custom designed. The next generation of plants should have much more predictable construction costs. Nuclear plant vendors claim that the next generation of nuclear plants (e.g., the Westinghouse AP-1000) should have capital costs close to competitive with baseload coal (around $2,000 per kW), but their ability to construct plants at this cost and on schedule has not been proven. New approaches to building nuclear reactors involve the use of the pebble-bed reactor, which is supposed to be safer, cheaper, and faster to build and roughly one tenth the size of the current generation of light water nuclear plants. New designs will help a new generation of nuclear power become more competitive. The South African pebble bed modular reactor (PBMR), the integral fast reactor (IFR), the high-temperature gas reactor (HTGR), and other ideas and designs are all likely to be considered in the future.

In addition, U.S. companies have gotten much better at operating nuclear power plants. The average utilization rate of nuclear plants has gone from around 60% in the 1970s and 1980s to well over 90% today. And the operating cost of nuclear plants (including fuel) has become less than one fourth of the operating cost of an average coal plant.

While some challenges to the future of nuclear power have been resolved, others persist. The cost of transporting nuclear waste remains high and the cost of decommissioning existing plants will similarly be high. The government has not yet completed plans for permanent storage of spent nuclear fuel, a cause for continued uncertainty. Building new plants will continue to be expensive because of high construction costs related to their being capital intensive and to the extended time required for construction. Community opposition to "siting" continues to complicate the issue. These factors will need to be resolved to make nuclear power economically attractive to new investors.

To the mix of variables impacting the economics of nuclear power one must add the role of "externalities." Production of electrical power regardless of source creates negative externalities for third parties not borne by the producer. These may be in the form of pollution, which has a possible negative health impact on people living near or downwind of plant, and the emission of greenhouse gases, which contribute to global warming. To calculate the "real" costs of producing electricity, one must factor in not only capital and operating costs but also the costs of all of these externalities, which can be difficult to calculate. For example, it is hard to estimate the cost of climate change from emission of greenhouse gases. Proposed carbon taxes in the future may increase costs for coal and gas producers, as will requirements for carbon capture and storage. Placing a price on CO_2 emissions using either an emissions tax or a cap-and-trade system could make nuclear power plants more competitive with coal plants. How much more competitive will depend on the level of charges for CO_2 emissions and the relative capital and fuel costs of alternative electricity generating technologies.

Use of nuclear power will increase in the long run only if it has a lower cost than competing technologies. Can nuclear power compete economically with coal and gas? In 2003 an interdisciplinary group of scientists, engineers, economists, and political scientists at MIT prepared a report entitled *The Future of Nuclear Power in the United States* (Joskow, 2006). In the report they examine the likely future economics of nuclear power and came to several conclusions. They start with a "base case," which is a calculation based on "real" costs for nuclear power and coal-, and gas-generating technology that uses assumptions investors would likely use to evaluate the costs of various alternatives. They conclude that under "base case" conditions—with current expectations about nuclear power plant construction costs, operating costs, and regulatory uncertainties—nuclear power would likely remain more expensive than gas or coal in the future. Nuclear power would likely be much more costly than coal in countries with good access to coal and without CO_2 emission charges. And prices for natural gas would be lower than for nuclear power too—as long as there are no CO_2 emission charges and as long as gas stays low. Higher gas prices would likely push investors to coal rather than nuclear power in regions where coal is available—as long as there were no CO_2 emission charges. Where coal is not available, higher gas prices would be likely to push investors to nuclear power.

The MIT report then developed several models that are based on various assumptions about construction costs, financing costs, fuel prices, and CO_2 emissions prices. One model assumes that regulatory, construction, and operating cost uncertainties can be resolved and that nuclear power plants can be financed at the cost of capital for coal or gas plants. Under this new set of circumstances, nuclear power then becomes more competitive with the costs of gas—and only a little more costly than coal.

A second model then factors in the impact of externalities or the "social costs" of carbon "taxes." These costs are the costs of carbon dioxide emissions and their projected impact on global climate change. Because nuclear power plants produce no carbon dioxide (unlike gas- or coal-fired plants), nuclear power becomes more attractive when the costs of carbon "taxes" are priced into the equation. In this model they use estimates ranging from carbon "taxes" (tC) at $50/tC to as high as $100/tC and $200/tC. The smaller amount is consistent with the Environmental Protection Agency's calculation of the cost of reducing U.S. CO_2 emissions by about 1 billion metric tons per year. With a lower carbon tax, nuclear power remains less economical in the "base case." With a higher carbon charge nuclear power can be competitive with coal—even without a reduction in construction costs. It is competitive with gas only if gas prices are high.

According to the MIT report, nuclear power can become competitive without CO_2 emission charges if construction costs—including the financing costs—fall by approximately 30–35%. This scenario is considered unlikely because of increasing competition for scarce people and material resources to construct the many new power plants being planned across the globe. More likely to make nuclear more competitive with both coal and gas would be a 25% reduction in construction costs and a higher carbon charge. This would be true in most places, except where gas prices are very low.

The MIT report concludes that using assumptions that commercial investors would be expected to use today in order to obtain an acceptable rate of return on their investment (not based on engineering estimates of what might occur under "ideal" conditions), the cost of nuclear power would exceed the costs of coal-fired plants and combined cycle natural-gas-powered plants even assuming a high price for natural gas. Nuclear power becomes competitive with coal or gas only in a model where construction costs are reduced by 25% and only when an average social cost that internalizes all relevant externalities is factored into the model.

In 2004, the University of Chicago issued a report entitled *The Economic Future of Nuclear Power*. In their financial model, they focus on the group of new nuclear power plants that will first come on line in the future. They suggest that with the benefit of the experiences of the first few nuclear plants, costs should become more competitive. In addition, they concluded that with time, the U.S. government is likely to place a higher priority on global warming and the development of alternatives to coal and gas-fired electricity. They projected that if stringent greenhouse gas policies are implemented in the future, the levelized cost of coal-fired and gas-fired electricity (LCOE) could increase, making nuclear power more competitive.

What Reporters Need to Know

The following conditions are among those likely to make nuclear power more economically viable in the future:

- A stable regulatory and licensing environment
- Financial incentives for those who build first
- Loan guarantees and tax credits for nuclear plants
- Significant "price" on carbon emissions from alternative fueled competitors to cover costs of externalities, including CO_2 emissions
- Resolution of nuclear waste disposal concerns

In the United States, new markets and institutions created under electricity deregulation (or restructuring) have made nuclear power plants highly profitable. Even still, the ability of new nuclear plants to compete with existing coal and natural gas plants is highly dependent on the supply/demand balance of these fossil fuels, as well as U.S. and multinational climate policy.

At the present time, U.S. policies regarding carbon emissions, caps and taxes, and targeted reductions remain uncertain.

The MIT and University of Chicago studies have unquestionable merit. However, the findings must be considered in the light of their use of certain assumptions and their use of specific numbers for alternative fuels. The MIT report makes assumptions regarding cost improvements that would make nuclear power plants more economical than gas or coal. As they note, the cost improvements that are projected are plausible but unproven.

Furthermore, the use of specific numbers means that as market conditions change, the conclusions can also change. Many of the numbers for alternative fuels used in these reports have become outdated. Natural gas prices have risen above the quoted $6/mmBTU figure necessary to make nuclear power competitive with natural gas. The cost necessary to get a new coal plant built has now surpassed $65 or $70 per MWh, about one and one half times the figure used in the reports.

New nuclear plants are costly and time consuming to build safely. According to a 2004 Brookings Institute policy brief (Nivola, 2004), the average costs are $3 billion (in 2002 dollars). Even if it took only 4 or 5 years to complete and, if it were possible to reduce construction costs by 25%, they conclude that nuclear power plants would still be more expensive to build than coal plants and just barely competitive with high-cost gas. Electricity produced by currently operating nuclear plants in the United States tends to be more cost competitive with gas- or coal-generated power once a plant has been paid off.

A fully depreciated nuclear power plant has costs far below coal and gas. Recent data from the Nuclear Energy Institute suggests that the operating costs of existing nuclear plants are less than $10/MWh, compared with $40 for coal. Between 1998 and 2002, more than a dozen older plants were sold, as large energy companies moved to use them as a hedge against increasingly unstable fuel prices for gas-fired generation, and in anticipation of possible additional environmental restrictions on coal-fired facilities.

There is still considerable perceived risk in the construction of new nuclear facilities. Government assistance in the form of tax credits or loan guarantees can help reduce this risk but likely will be limited to the first few of each type of next-generation plant. If nuclear operators demonstrate consistency in building new nuclear plants, this risk should drop. Public perception issues and waste management problems, however, are likely to persist. And while it is true that the risks of new power plant construction are now on the shoulders of investors rather than ratepayers, investors will likely not shy away from nuclear plants simply because they have long lead times and are capital intensive. The natural gas bubble in the United States is proof. At the onset of industry deregulation, investors flocked to new gas-fired projects. But the ensuing increase in natural gas prices meant that in a competitive market, these plants were not economical. The plants were not dispatched, and the merchant generators have taken large losses. Many of these merchant gas plants are being sold off at fire-sale prices. In the United States, new nuclear plants are being considered largely by consortia of nuclear operators—a mix of utilities and merchant generators. These shared ownership arrangements are another way to address some of the risks involved in constructing new nuclear plants.

Pitfalls

A renewed interest in the use of nuclear energy to curb carbon dioxide emissions will likely generate new economic concerns. Building nuclear power plants may reduce a country's reliance on imported oil and gas, adding to security in one way but also generating new targets for terrorist attack and increasing amounts of enriched uranium that must be protected from the danger of possibly being diverted to make nuclear weapons. (As noted elsewhere, it would be a massive economic, scientific, and engineering challenge to convert used fuel to a weapons grade material, and proliferation nations have not used that path.) Policing nuclear technology and managing risks will become an even greater challenge.

In a competitive environment, nuclear power plants will have to compete on the basis of cost, without large government subsidies. Nuclear power, however, continues to receive subsidies from the government—as do other fuel sources—in the form of taxpayer-funded research and development and limitations on disaster liability. For example, with the new generation of power plants being planned, the U.S. government plans to offer the first 6,000 Mw subsidies equal to those offered to renewable sources and some compensation for possible cost overruns due to regulatory delays.

Programs established under the Federal Energy Policy Act, while making it easier to obtain long-term funds from the bond markets, will not appreciably reduce the cost of construction and the large front-end investment nuclear power plants require.

The nuclear industry has put forward optimistic estimates of new construction costs that are very uncertain. Lack of recent nuclear plant construction makes it hard to estimate actual construction costs. Estimates by the industry reflect "best case" assumptions. While these estimates contain an element of uncertainty, one thing is certain. No one should ever underestimate the costs of a nuclear power plant during the pre-construction phase. The nuclear plant being built by Areva and Siemens in Finland has already experienced cost overruns that are estimated to be about 20% of the original contract price and delays in the anticipated operating date.

It is true that there has been a reduction in licensing barriers; however, the new licensing process has never actually been tested. No one knows how it will work. This is a major "first of the kind" uncertainty and this has become the rationale for the "insurance" provided in the 2005 Energy Policy Act.

In countries with direct access to low-cost fossil fuels, nuclear power is less likely to become cost competitive with other sources of electricity generation without CO_2 taxes or limitations.

While often forgotten in calculating costs, the costs of waste disposal, infrastructure requirements, environmental impact, and decommissioning need to be considered when comparing the cost competitiveness of various alternate energy sources. These costs are included in the cost estimates for electrical generation using nuclear power.

Part of calculating the economics of nuclear power involves calculating the costs of externalities. Part of this calculation are the costs of carbon capture and storage (CCS), which is hampered by the fact that technologies are still under development and the science of what happens to sequestered carbon dioxide is incomplete. While the disposal of carbon dioxide is straightforward, the technology for pulling it out of emissions is not. Technically carbon capture, storage, and sequestration are possible; making it commercially viable from a cost point will take time. The safety implications of CCS have not been fully investigated, and large scale releases of stored CO_2 from relatively frequent geological events (recall the African lake event that killed over 1,500 people) could lead to much higher mortalities than might be acceptable.

Putting the option of future expansion of nuclear power back on the table does not necessarily mean nuclear energy will replace fossil fuels any time soon. In many ways, revival of interest in nuclear power illustrates the lack of acceptable choices for combating global warming. Renewable energy still has its limitations. The wind must blow to turn windmills, solar power and geothermal energy are still not economically viable, and hydroelectric dams disrupt communities and habitats.

Expansion of nuclear energy as a substitute for coal-produced energy to curb carbon dioxide emissions would require a large increase in the number of nuclear power plants in operation—according to some estimates a 10-fold increase to make it capable of generating one third of the world's electricity.

References and Other Resources

The economic future of nuclear power: A study conducted at the University of Chicago. (2004, August). *www.ne.doe.gov/np2010/reports/NuclIndustryStudy-Summary.pdf.*

The future of nuclear power: An interdisciplinary Massachusetts Institute of Technology study. (2003). *web.mit.edu/nuclearpower.*

Joskow, P. L. (2006, December). *The future of nuclear power in the United States: Economic and regulatory challenges.* AEI-Brookings Joint Center for Regulatory Studies. Working paper 06-25. *envirovaluation.org/index.php/2006/12/27/aei_brookings_joint_center_www_aei_brook_30.*

Mazuzan, G. T., & Walker, J. S. (1985). *Controlling the atom: The beginnings of nuclear regulation, 1946–1962.* Berkeley: University of California Press.

Nivola, P. (2004). *The political economy of nuclear energy in the United States.* Policy brief 138. Washington, DC: The Brookings Institute. *www.brookings.edu/comm/policybriefs/pb138.htm.*

Walker, J. S. (2000). *A short history of nuclear regulation, 1946–1999.* Washington, DC: U.S. Nuclear Regulatory Commission.

World Nuclear Association. *World nuclear power reactors 2007–08 and uranium requirements.* (Site also provides links to further information.) *www.world-nuclear.org/info/reactors.html.*

World Nuclear Association. (2007, June). *The economics of nuclear power.* Briefing paper 8. *www.uic.com.au/nip08.htm.*

Civilian Uses of Radiation and Radioactive Material (Other than Commercial Nuclear Power)

Written by Bernadette M. West, based in part on an interview with P. Andrew Karam, with comments by David C. Kocher

Background

We often fail to recognize the many beneficial civilian uses of nuclear materials other than commercial nuclear power. These include use of nuclear materials used in the field of medicine and in food preparation and a variety of industrial applications. The U.S. Nuclear Regulatory Commission (NRC) is the agency that has authority over civilian uses of nuclear materials in the United States.

Identifying the Issues

Diagnostic Nuclear Medicine

In the field of nuclear medicine health professionals conduct diagnostic tests that create images of the body that can be used to identify and stage certain diseases, such as cancer. These tests detect gamma rays emitted from a radioactive substance given to the patient either orally or intravenously to create an image of what is happening in the body.

The type of radioactive tracer used depends on the target organ. The tracer accumulates in the organ and gives off energy as gamma rays that a special crystal within a gamma camera is able to detect. Patients lie on a scanning table and a specialized nuclear imaging camera is used. Pictures and measurements of the organ and surrounding tissue are taken. These images can be used to help diagnose tumors, infection, and other disorders. The procedures are usually time consuming and can take several hours to perform.

The findings from nuclear medicine are often critical in diagnosing and treating certain diseases. These tests provide information about the functioning of an organ within a specific region of the body that often cannot be obtained from other diagnostic tests, such as x-rays. Nuclear medicine tests can help diagnose many conditions, including cancer, infection, arthritis, bone fractures, and blockages of the gallbladder, and assess blood flow, thyroid and kidney function, and functioning of the heart and lungs.

Positron emission tomography (PET) is an example of a nuclear medicine test. A PET scan is a diagnostic test that produces a three-dimensional image or map showing the functioning of parts of the body. In a PET scan, the patient is injected with a radioactive substance that emits positrons, which are positively charged electrons. These positron emitters (the PE in PET) are attached to chemicals or drugs that travel to a specific part of the body, and then a scan is done. When the emitted positrons come into contact with nearby electrons, gamma radiation is produced that can then be detected by the scanner and an image created. On a PET scan, different colors and levels of brightness can be observed. Cancerous tissue will appear brighter than normal tissue on the PET images.

The images produced by the PET scan are used to diagnose cancer and to calculate the effects of cancer therapy. They are also used to observe blood flow to the heart and to determine the status of heart tissue following a heart attack. They are also used in studies of the brain involving memory disorders and to look for possible tumors or to explain seizure disorders. The PET scan helps health professionals detect changes in biochemical processes that suggest disease before changes in anatomy are apparent with other imaging tests, such as the CT scan (computed axial tomography) or MRI (magnetic resonance imaging). The radioactivity is very short-lived. Radiation exposure is low and the substance amount is so small that it does not affect normal processes of the body.

Nuclear Medicine and the Treatment of Disease

In addition to its use in diagnosing disease, radioactive substances are sometimes used in treating disease and for palliative purposes. While many of these treatments involve low doses and minimal risk, it is important to note that some treatments using radioactive substances use higher doses and involve higher levels of risk.

Nuclear medicine is used to deliver chemotherapy or other treatments to the exact location where they are needed, then allowing health care providers to monitor how the body is responding to the treatment. Medicines with trace amounts of a radioactive material called radiopharmaceuticals are used in nuclear medicine. For example, iodine-131 is used to treat thyroid cancer and other diseases of the thyroid. In cases of thyroid cancer, radioactive iodine is used after surgery to eliminate any remaining thyroid tissue. A nuclear scan following the treatment for thyroid cancer also allows the physician to determine whether the radioactive iodine has gone to other parts of the body, indicating possible spread of the thyroid cancer.

Radioactive substances are also used to help reduce the pain from some forms of bone cancer. Patients with cancer that has spread into the skeletal system may experience much pain. Samarium-153 or strontium-89—radiopharmaceuticals that are injected into the bloodstream and travel to cancer lesions in the bones—are used as pain control therapies.

Food Irradiation

The first uses of food irradiation were begun in the 1950s. Today most spices, some fruits, and a small number of meat products sold in stores are irradiated. The food industry uses radiation to kill germs such as E. coli or salmonella that could make people sick. Irradiation is used to destroy insects and parasites in food products such as grains, dried beans, dried fruits and vegetables, and meat and seafood. It is also used to keep crops such as potatoes and onions from sprouting. It can be used to delay ripening in fresh fruit and vegetables. Not only does radiation kill germs, it also extends the shelf life of food because it also kills germs that cause food to spoil more quickly.

When food products are irradiated, the radiation does not stay in the food. The radiation kills dangerous organisms, such as bacteria, but it is not retained in the food. Studies show that when food is irradiated, there are small losses of nutritional value. For example, some foods may lose some vitamins but the amount lost is not considered significant in relation to the entire diet. Any kind of treatment of food (cooking, canning, or irradiation) causes some minor chemical reactions within the food. Heating creates thermolytic products; irradiation creates radiolytic products. The chemical changes from food irradiation are considered much less than comparable changes caused by cooking or canning foods. Thermolytic changes are so significant one can actually smell and taste them—as when foods are fried or grilled.

In the irradiation process, the food is exposed to high levels of gamma radiation. Cobalt 60 is the radionuclide that is used to irradiate food using gamma rays. Cobalt 60 is made by bombarding the metal cobalt 59 with neutrons in a nuclear reactor. The finished product is then encapsulated in stainless steel tubes shaped like pencils so it does not leak. Cobalt 60 has a half-life of 5.3 years. Most cobalt 60 comes from Canada.

The gamma rays penetrate the food to a depth of several feet. The irradiation process occurs in a chamber shielded by thick concrete walls to protect workers from the gamma rays. To enter the chamber, workers must pass through a labyrinth that provides shielding of radiation and prevents radiation from streaming out through the opening into the adjoining room. When the radioactive "source" is not being used, it is stored in a pool of water that absorbs the radiation harmlessly and completely. When food is irradiated, the source is pulled up out of the water into a chamber with massive concrete walls that keep any radiation from escaping. The food that is being irradiated is brought into the chamber and exposed to the radiation for a specified period. Then the source is lowered back into the pool of water. This technology has been in use for over 30 years.

There are three dose categories when foods are irradiated with gamma rays:

- Low-dose irradiation (up to 1 kGy) inhibits sprouting, delays ripening, eliminates infestations of insects, and kills parasites.
- Medium-dose irradiation (1 to 10 kGy) reduces microorganisms and non-spore-forming pathogens (i.e., disease causing microorganisms) that cause foods to spoil.
- High-dose irradiation (>10 kGy) reduces microorganisms to the point of sterility.

As the level of contamination increases, the level of irradiation needed to eliminate bacterial contamination increases. And when foods are highly contaminated, the higher doses of irradiation that become necessary could impact taste and texture.

In addition to the use of gamma rays, foods are also irradiated using electron beams (e-beams) generators and x-ray accelerators. The e-beam is a stream of high-energy electrons. These electrons are propelled from an electron gun that can be switched on or off. No radioactivity is involved. Some shielding to protect workers from the e-beams is necessary. This technology allows electrons to penetrate food only to a shallow depth of about an inch. Use of e-beams offers certain advantages: they can be turned on only as needed and there is no radioactive waste. But they can penetrate only shallow depths and they use large amounts of electricity. For thicker food substances, such as carcasses, gamma radiation with greater penetration or x-rays may be more appropriate.

X-ray acceleration involves the use of a powerful machine that produces x-rays using a beam of electrons directed at a plate of gold or other metal. A stream of x-rays coming out the other side can penetrate foods. X-rays are the same type of radiation as gamma rays, the only difference (other than in how they are produced) is that x-rays have lower energies and, therefore, somewhat smaller penetration depths in food. Like gamma rays, x-rays require heavy shielding for worker safety. Like e-beams, the machine can be switched on and off, and no radioactive substances are involved.

Since the mid-1980s irradiated products must be stamped with the international logo (the *radura*), which looks similar to a flower. The FDA requires that both the logo and the phrase "Treated with irradiation" be stamped on packages with food that has been irradiated or placed near locations within stores where fresh products are sold that have been irradiated. Individual food processors may add information explaining why the food was irradiated, such as "treated with irradiation to inhibit spoilage."

Industrial Applications

Many industries use radioactive materials. A number of industrial processes involve devices or facilities called irradiators in which products are exposed to radiation in order to sterilize them. This includes food irradiators, but also irradiators that sterilize milk containers and hospital supplies.

Gauges containing sealed sources of radiation that radiate through substances are used to measure, monitor, and control the thickness of materials, such as sheet metal, textiles, paper napkins, newspaper, plastics, photographic film, and other products as they are being produced. Knowing that radiation loses energy as it passes through a substance, the industry has used radiation to develop very sensitive gauges to measure thickness and density of materials. Imaging devices have been developed that can be used to look for weaknesses and flaws in finished goods.

There are numerous examples of industrial uses of radiation. The steel industry uses radioactive materials to test the quality of sheets of steel being produced. It is used to look for cracks in steel and to measure for uniformity of thickness. Similarly, automobile manufacturers use radioactive materials to test the steel they use in cars. Can and bottle manufacturers use radioactive materials to obtain the proper thickness of tin, aluminum, and glass. The mining industry uses radionuclides to help find mineral deposits. Special nuclear material tracers are used to help explore for oil, gas, or minerals in deep wells. This process is referred to as well-logging. When building new roads, construction crews use radioactive materials to gauge the density of road surfaces and subsurfaces. Small amounts of a radioactive substance are sometimes used to track leaks from piping systems and to monitor engine wear and corrosion of processing equipment.

Radiation is used in industry to produce images of the insides of tanks in order to inspect parts made of metal and welds for defects. Use of radiation in level gauges makes it possible to take indirect measurements in tanks where heat or corrosive substances make it impossible to take direct measurements with gauges. For example, in measuring a corrosive substance in a tank, level gauges containing radioactive sources can be used that alert workers to tanks approaching full capacity or tanks in danger of reaching empty.

Other Civilian Uses

Research is another important area that often relies on radiation and use of radioactive materials. Radiation and radioactive materials have been essential to all kinds of major advances in biological and physical sciences. Academic institutions may use nuclear materials in classroom demonstrations and laboratory research and to provide health physics support to other institutional nuclear materials users.

Finally, small amounts of radioactive materials are sometimes used in consumer products that provide great benefit to the public. For example, Am-241 is used in smoke detectors. Another familiar consumer product is timepieces that contain small amounts of tritium (H-3).

What Reporters Need to Know

There are three major types of radiation emanating from a radioactive nucleus: alpha (α), beta (β), and gamma (γ). Each type of radiation has slightly different effects on a biological cell, described as the "relative biological effectiveness" (RBE) of that type of radiation at a specified energy. All radiation acts by breaking molecular bonds and creating free radicals. In diagnostic nuclear medical applications, patients receive a radiation dose similar to that resulting from a standard x-ray. In nuclear medicine applications, most radiation goes into the patient. Most of the radioactivity then passes out of the body in urine or stool and the rest disappears through natural loss of radioactivity that occurs gradually. These tests are less traumatic than invasive exploratory surgeries. Allergic reactions can occur but they are extremely rare. The dose of radiation delivered to the patient is very small and studies indicate that the risks are very low (e.g., much lower than risks that people experience and accept in everyday life, or much lower than risks from natural background radiation). As with any test involving radiation, exposure during pregnancy should be kept to a minimum. In certain tests, radioactive materials are given that would require a nursing mother to stop breastfeeding.

There are no known health risks from eating irradiated food. While food irradiation is used to make food safer, it is not meant to be a substitute for sanitary conditions on farms or stringent food processing procedures. Irradiation does not eliminate the need for safe food handling procedures in the home. Irradiated foods can be recontaminated by bacteria found in the refrigerator or on kitchen counters. Food irradiation also cannot be used to make spoiled foods good or to "clean-up" dirty foods. It cannot reverse the effects of spoilage.

Irradiation is referred to as a cold process. The process does not significantly increase the temperature of a food or change its characteristics. For example, when an apple is irradiated, it remains crisp and juicy. When meat is irradiated, it is not cooked. Certain foods cannot be irradiated because the process causes them to change flavors, as with dairy products, or to change consistency, as with peaches or nectarines.

Radiation used to kill bacterial contamination (which does not cook food) is different from microwave radiation (which does cook food).

Many of the commercial uses of radioactive materials discussed earlier (diagnostic and therapeutic nuclear medicine, food irradiation, industrial processes, and medical research) generate some amount of low-level waste. Radioactively contaminated industrial or research waste includes such things as paper, rags, plastic bags, packaging materials, protective clothing, organic fluids, and water-treatment resins. It is generated by industries and institutional facilities (e.g., universities and hospitals). In nuclear medicine, most of the radioactive material decays within the patient's body or is excreted—so there is no waste from the administration of radioactive materials that then requires disposal, although there may be waste requiring disposal from preparation of the radioactive material that is administered to a patient. Low-level waste is less dangerous than high-level waste. It is shipped to low-level waste disposal facilities where it is packed, buried in trenches, and covered with soil. Although states use already existing low-level waste disposal sites, many will have no place to dispose of low-level waste generated in the civilian sector when the facility at Barnwell, South Carolina, stops accepting waste from most other states. And there are no new disposal facilities for civilian low-level waste under development at present.

Pitfalls

People often confuse PET scans with computed tomography or CT scans. There are important differences between the two. A PET scan and a CT scan are both tests used to diagnose disease. There are, however, important differences (see Table 2).

The discussion on low doses and risks from diagnostic nuclear medicine procedures does not apply to all therapeutic procedures. It is important to note that some therapeutic procedures use radioactive substances at higher doses and involve higher levels of risk.

Although organic food is labeled nonirradiated, irradiation does not make food nonorganic.

Table 2. A Comparison of CT Scan and PET Scan as Diagnostic Tools

CT scan	*PET scan*
It uses x-rays to take pictures of structures in the body.	It uses an injection of radioactive material to assess the function and processing of materials by organs.
It is an x-ray generating machine; it sends a narrow beam of x-rays through the body to generate an image.	It involves injecting (or swallowing) radioactive materials, which travel through the body to the organ of concern. "Gamma cameras" are then used to see where the radioactive materials are.
The source of radiation is outside the body.	The source of radiation is on the inside—in the form of radioactive drugs.
A CT machine can be turned off.	The radioactivity from radioactive materials cannot be turned off; it decays or is excreted from the body.
It uses higher levels of radiation, given in a shorter period.	It produces lower levels of radiation, given over a longer period.
It does not make the recipient radioactive.	Ingesting nuclear medicine nuclides will make the recipient temporarily radioactive (although not dangerously so).
It shows the doctor what the structure (organs or bones) looks like but provides no information about the inner workings.	It allows the doctor to see the process of metabolism (i.e., where the radioactive materials are being used within the body, such as the growing surface of a bone).

Note: The information in this chart about doses and risks does not apply to all therapeutic procedures in nuclear medicine.

Resources

American Council on Science and Health home page. *www.acsh.org.*

Centers for Disease Control. *Food irradiation. www.cdc.gov/ncidod/dbmd/diseaseinfo/foodirradiation.htm#whatis.*

Health Physics Society home page. *hps.org.*

Idaho State University. Radiation Information Network's *Food irradiation. www.physics.isu.edu/radinf/food.htm.*

Irradiation. Federal, state, and international Web sites for food safety information. *www.foodsafety.gov/~fsg/irradiat.html.*

Society of Nuclear Medicine home page. *www.snm.org.*

U.S. Food and Drug Administration. *Food irradiation: A safe measure. www.fda.gov/opacom/catalog/irradbro.html.*

U.S. Nuclear Regulatory Commission. *Industrial uses of nuclear materials. www.nrc.gov/materials/miau/industrial.html.*

Section 3

Nuclear Waste Management

Nuclear Waste Policy in the United States: Classification, Management, and Disposition

Written by Karen W. Lowrie and Michael R. Greenberg
with Richard B. Stewart and Jane B. Stewart, based in
part on interviews with Frank L. Parker and James Bresee,
with comments by Keith Florig and Buzz Savage

Background

Nuclear waste is defined as liquid, solid, or semi-solid waste products possessing at least some amount of radioactive elements. It consists of radioactive materials left over from some industrial, scientific, military, or medical process that uses radiation. The more radioactive the waste is, the more stringent the regulations for how it is stored, transported, and collected at disposal sites. This chapter describes the different legal classifications of nuclear waste in the United States, the regulations applying to each of those classifications, and the challenge of finding sites for permanent disposal of nuclear waste.

There are two main streams of nuclear waste in the United States: civilian and military. Civilian nuclear waste consists of spent fuel from commercial nuclear reactors and other radioactive waste generated from medical use or other industries. The overwhelming majority of the total amount of radioactivity in all U.S. nuclear waste is in the spent fuel rods from nuclear power generation. The long rods are arrayed in square bundles whose physical appearance does not change while the fuel is in the reactor. After the uranium fuel has been in the reactor for several months, its ability to produce energy from fission begins to decline. After about 18 months, the fuel must be replaced with fresh fuel. When the spent fuel is removed from the reactor, it is highly radioactive or "hot" and must be handled remotely.

Although the majority of radioactivity is in the spent fuel rods, the majority of the total volume of nuclear waste is from the defense sources. The defense waste, produced from decades of nuclear weapons production and research, is not as "hot" as the commercial civilian waste because it did not stay in reactors for a long time. Very little new high-level waste has been generated by defense sources since the United States ceased producing nuclear weapons. Yet there is a legacy, and the U.S. Department of Energy (DOE) is processing a substantial amount of defense-related waste at the Savannah River, South Carolina, site as part of the cleanup. Because radioactive materials are continually decaying, their concentrations are now much lower in these wastes than when they were originally produced. Another major source of waste are nuclear fuel cycle activities to support these industries, such as uranium mining and processing.

Classification of Wastes

Waste is classified by federal laws, regulations, and rules, and this classification has evolved during the history of nuclear material uses and does not necessarily correspond to hazard levels. For example, some low-level waste could contain highly radioactive items. About 90% of radioactive waste, by volume, is low level, about 10% is intermediate level (see note following list of classifications), and about 0.3% is high level. High-level waste, however, contains 95% of the total radioactivity of all nuclear waste.

The main classifications are as follows:

- *High-level waste (HLW).* These wastes are the byproduct of the reactions that occur inside nuclear reactors. The U.S. Nuclear Regulatory Commission (NRC) defines high-level waste as spent fuel rods and reprocessing waste. Reprocessing consists of extracting isotopes from spent fuel that can be used again as reactor fuel. The reprocessed waste is liquid and solid waste from solvent extraction cycles. In addition, quantities of high-level radioactive waste are produced by the defense programs at the DOE. These wastes, which are generally managed by the DOE, are not regulated by NRC.
- *Transuranic waste (TRUW).* This is waste, regardless of origin, containing elements with atomic numbers (number of protons) greater than 92, the atomic number of uranium. (Thus the term *transuranic,* or "above uranium.") Most of this waste was produced from defense-related activities. Transuranic waste includes only waste material that contains transuranic elements with half-lives greater than 20 years and concentrations greater than 100 nanocuries per gram. If the concentrations of the half-lives are below the limits, it is possible for waste to have transuranic elements but not be classified as transuranic waste. At present the United States permanently

disposes of transuranic waste generated from nuclear power plants and military facilities at the Waste Isolation Pilot Plant in New Mexico.

- *Low-level waste (LLW).* This waste is defined by what it is not, that is, it is radioactive waste not classified as high-level, spent fuel, transuranic, or byproduct material, such as uranium mill tailings (see next bullet item). It thus includes everything from slightly radioactive trash (such as mops, tools, syringes, and protective gloves and shoe covers) to highly radioactive activated metals from inside nuclear reactors. It is produced at thousands of locations, including government agencies, industries, power plants, research universities and hospitals. It consists of mostly short-lived but also some long-lived radionuclides. Some low-level waste could contain some highly radioactive items. Low-level waste is typically stored on-site by licensees, either until it has decayed away and can be disposed of as ordinary trash, or until amounts are large enough for shipment to a low-level waste disposal site in containers approved by the U.S. Department of Transportation. Low-level waste has four subcategories according to activity level and life-span: classes A, B, C, and greater than class C (GTCC). On average, class A is the least hazardous, while GTCC is the most hazardous.
- *Uranium milling residues.* Uranium mine tailings are soil, rock, and other by-products left over after uranium has been extracted from ore. (See "Sustainability.")
- *Mixed wastes.* In both the commercial and the military sector, some of the radioactive wastes generated are mixed with hazardous substances, such as organic solvents or other toxic chemicals. Much of this waste (especially the transuranic waste) contains substantial quantities of long-lived radionuclides, such as plutonium-239 and technetium-99. The radioactive components of mixed wastes are regulated under the Atomic Energy Act by the NRC for commercial sources, and by the DOE for defense sources. The hazardous components, however, are subject to regulation by the Environmental Protection Agency according to an environmental law known as the Resource Conservation and Recovery Act.

(*Note:* Intermediate-level waste, consisting of resins, sludge, and other materials, contains a higher level of radioactivity than low-level waste. However, U.S. regulations do not define this category of waste; the term is used in Europe and elsewhere.)

National Nuclear Waste Policy and Regulatory Agencies

In U.S. history, nuclear energy and defense policies were developed like "buildings built without toilets." In other words, production of power and production of weapons was the paramount concern, but little or no attention was paid to the waste disposal issues. The Atomic Energy Act, which provided the framework for the early production of nuclear power in this country, had limited provisions for dealing with waste.

The nuclear wastes from civilian nuclear energy programs of greatest concern are (1) spent nuclear fuel rods used to generate electricity in reactors (civilian use) and (2) high-level wastes produced from chemical reprocessing of irradiated fuel rods to extract uranium and plutonium for making nuclear weapons used in World War II or stockpiled bombs. Since many of the spent civilian fuel rods will be radioactive for hundreds of thousands of years, they will require isolation from the human environment for a period far longer than all of recorded history, unless, as noted earlier, actions are taken to transmute highly radioactive elements into less radiotoxic isotopes.

The historical assumption was that the rods would also be reprocessed when they reached the end of their useful life, extracting uranium and plutonium as new fuel. This measure would postpone the waste problem and reduce the overall volume of nuclear waste generated, although it would produce high-level waste. A small amount of reprocessing of spent civilian nuclear fuel did in fact occur in the late 1960s and early 1970s. The West Valley reprocessing facility was built in New York State and operated for 6 years (closed in 1972). Two other reprocessing facilities were built but not opened. In the 1970s, President Jimmy Carter issued an executive order to prevent reprocessing in the United States, out of concern for proliferation of the separated plutonium for weapons use and in the hope that his action would spur other nations to similarly ban reprocessing. (See "Nuclear Nonpoliferation.") His issuing the order meant that the spent fuel waste problem would remain. Spent fuel from nuclear power plants became waste and commercial reprocessing ended. President Ronald Reagan lifted the ban on reprocessing but by then the industry did not believe reprocessing could be performed economically.

In 1982, Congress passed the Nuclear Waste Policy Act (NWPA, P.L. 97-425) to deal with the issue of spent fuel disposition. It called for the disposal of spent nuclear fuel rods and high-level waste in a "deep geologic repository." The policy requires encapsulation of concentrated process wastes from defense sites into a glass matrix (a process called vitrification). Spent fuel is to be loaded into canisters.

The legislation specified roles for three different agencies: the DOE, the NRC, and the U.S. Environmental Protection Agency (EPA). The DOE was required to create the Office of Civilian Radioactive Waste Management, designated to be in charge of locating sites for permanent geologic disposal of spent fuel and high-level wastes. The NRC was required to regulate the design and operation requirements of the repository, including such things as how the waste would be packaged and stored. The EPA would issue environmental standards to protect humans and the environment from the release of radiation from the repository and would be responsible for assessing the security and protectiveness of the site.

The NWPA set up a process that required the DOE to look for at least two suitable sites; it was expected that there would be one site in the western United States and another in the eastern United States. In the first round of siting, the DOE undertook the process of finding a site in the west. Nine potential sites were identified; then these were winnowed down to five potentially suitable sites. Once these five sites were screened, the list was further narrowed to three candidate sites, one in southeastern Washington, one in Deaf Smith County, Texas, and one about 90 miles north of Las Vegas in southern Nevada (Yucca Mountain). A second round of siting was also begun, and it screened several potential repository sites in the eastern United States. (See MacFarlane and Ewing, 2006, for an excellent treatment of the Yucca issue.)

Under the political leadership of powerful members from Washington and Texas, Congress amended the NWPA in 1987 to say that Yucca Mountain in Nevada would be the sole nuclear waste repository, in essence stopping the prior two-round selection process. This decision was defended on the ground that Yucca was a good site in terms of geology, weather, and population density, and that it would be very costly to conduct further detailed studies of additional sites. Five years later, in the Energy Policy Act of 1992 (P.L. 102-486), Congress directed the NRC and the EPA to set regulations for design, construction, and operational and environmental standards specific to the site and circumstances of Yucca Mountain. Transportation of wastes to Yucca was to be regulated by the Department of Transportation and by the NRC. Critics of this "Yucca only" policy say that creating site-specific standards means that the standards will be more lenient and easier to meet than if the site had to meet the general regulations that would apply to any site. Supporters say, however, that it makes sense to customize standards to the site. Since 1992, the NRC and the EPA have promulgated regulations specific to Yucca Mountain. These standards were successfully challenged in court and are being revised. (See the discussion in the next section, "Identifying the Issues.") The current EPA-proposed Yucca rule, which when finalized will be tracked by NRC's Yucca rule, would set a 15-millirem annual dose limit for the "reasonably, maximally exposed individual" for the first 10,000 years of repository operation; the annual dose limit for the period from after 10,000 years to 1 million years would be 350 millirem. As of 2008, the EPA has not yet finalized this rule.

The second repository that was to be located in the eastern United States was politically opposed and the DOE did not conduct further studies of any additional sites, in compliance with congressional direction. Since the late 1980s, the state of Nevada has consistently protested the "Yucca only" decision and has unsuccessfully sued the federal government several times to stop it. Opponents have raised questions, for example, about what we know about the suitability of this site, the stability of the geologic formations, the likelihood of earthquakes, possible water infiltration, and possible human intrusion. But the law provides for a recommendation from the secretary of energy to the President to go forward with Yucca, and Energy Secretary Spencer Abraham did so in February 2002, along with submission of a final environmental impact statement (EIS). The law also provided the state of Nevada the opportunity to reject this recommendation through a "notice of disapproval," which is binding unless Congress approves the site as a repository and the President signs that approval into law within 90 days of the notice's receipt. But in July 2002, after many years of scientific studies and political debates, Congress overrode Nevada's opposition and the President signed the law reaffirming Yucca as the waste repository.

Thus, Yucca Mountain is now the only candidate site for a permanent repository for spent nuclear fuel and high-level wastes. A repository at Yucca Mountain must still be licensed by the NRC, which among other matters will require the NRC to comply with EPA radiation release standards. The Office of Civilian Radioactive Waste Management (OCRWM) is the NRC license applicant for Yucca Mountain. The OCRWM submitted a license application to the NRC on June 3, 2008. The NRC docketed the application in September 2008 and recommended adoption of the DOE's Environmental Impact Statement for the project, subject to receiving some supplemental information. Congress set a 3-year deadline after docketing for the NRC to make a construction authorization decision. OCRWM estimates that the Yucca Mountain facility will be licensed by 2010 at the earliest and opened by 2017 at the earliest. In reality, it may be much longer because of continued opposition and the likelihood of continued legal challenges. With Senator Harry Reid's rise to majority leader in the U.S. Senate, there may be political problems with the Yucca Mountain project's moving forward. Senator Reid has already declared that the Yucca Mountain facility is "never going to open" (Reid, 2007).

Interestingly, the disposal of transuranic waste has been handled more successfully. Transuranic waste generated by the federal government in the process of making nuclear weapons is being deposited and stored in a salt dome on federal land at the Waste Isolation Pilot Plant (WIPP) near Carlsbad, New Mexico, pursuant to the DOE's general authorities under the Atomic Energy

Act. The site was first identified in the mid-1970s and began receiving wastes in 1999. It has operated successfully since then. WIPP has rules on what it can and cannot accept. About 97% by volume is stored in drums and does not need to be handled remotely. The remainder must be stored in casks shielded with concrete, lead, or steel to block the gamma radiation, and it requires remote handling.

New Mexico successfully litigated to establish the state's right to be involved in all significant decisions relating to WIPP. The state's concerns were eventually resolved, in part through federal funding for state measures to ensure safe transportation of wastes to the site. Congress passed legislation in 1992 to approve use of federal lands for the site. Some observers assert that it was easier to site the transuranic waste facility at WIPP for these reasons: the waste is solely defense waste rather than spent commercial nuclear fuel, the local community has strongly supported the facility, and New Mexico has been a more willing participant, ready to work out issues.

While spent nuclear fuel, high-level waste, and transuranic waste are a federal responsibility, states and commercial disposal facilities have primary responsibility for handling low-level commercial wastes. Typically, low-level waste is retained on site by licensees until it is has decayed to the point of posing no measurable risk and therefore can be disposed of as trash, or it can be stored in containers and then shipped to a commercial low-level radioactive waste management site. These privately owned facilities are located in South Carolina, Utah, and Oregon. These facilities are licensed by the NRC, or by the state in "agreement states" that are approved by the NRC. Under the Low-Level Waste Policy Act of 1980, as amended in 1985, groups of states have entered into interstate agreements to site new facilities and manage the group's wastes at a site in one of the group member states. But none of the ten approved group state disposal compacts, or individual states that have not joined compacts, has succeeded in finding a new low-level waste disposal site. The 1985 amendments allow facilities to reject waste from states that are not part of their regional compact. When the three operating facilities are filled to capacity or not accepting new waste, the issue of finding additional suitable low-level waste storage sites will come to the fore again.

Another piece of federal policy is transportation of spent fuel or other nuclear wastes. During the past 30 years, thousands of shipments of spent fuel and high-level waste have crisscrossed the United States. However, if the Bush administration's proposed Global Nuclear Energy Partnership is approved, if national interim storage sites are established, or if any other changes are made that require moving spent fuel, then the number of shipments would substantially increase and transportation would become a major policy issue for local and state governments. (See "Closing the Civilian Nuclear Fuel Cycle.") Notably, a survey of people who lived within 50 miles of six DOE former weapons sites showed almost as much concern about shipments as about on-site processing and management of wastes (Greenberg et al., 2007a). Furthermore, there is always the chance, no matter how remote, of a terrorist attack on a nuclear waste shipment. Hence, along with the NRC and the DOE, the U.S. Department of Transportation will be called on to be intimately involved with any nuclear waste issue.

Finally, the issue of reprocessing has again arisen. The proposed Global Nuclear Energy Partnership is a new program to reprocess spent fuel rods and reuse the uranium, plutonium, and other transuranics for fuel. The process would also produce radioactive wastes. Some of the uranium fuel would be exported to other countries as part of a strategy to prevent proliferation of nuclear weapons. In other words, countries could have nuclear power without establishing their own uranium enrichment or reprocessing facilities, which could be used to separate materials for nuclear weapons. The plutonium could be burned as fuel in existing reactors or in a new type of reactor to be developed and built in the United States. If it succeeded, the reprocessing and recycling would reduce the total volume and radiotoxicity of highly radioactive wastes, reducing the amounts that would require geologic disposal. But the program is controversial and has been challenged on economic, technological, and other grounds.

Identifying the Issues

The primary debates, conflicts, and issues about nuclear waste policy are focused on the question of spent nuclear fuel and high-level waste disposition and the Yucca facility itself. Key issues are the following:

- The license to actually construct the Yucca facility has not been issued, and EPA radiation release standards have not yet been settled. The EPA issued protective standards, but in 2004 they were successfully challenged in a suit brought by the state of Nevada, the Natural Resources Defense Council, and other plaintiffs in the U.S. Court of Appeals. The 1992 act directed the EPA to follow a report issued by the National Academy of Sciences on environmental standards for Yucca. The report recommended that standards should be set for longer than the 10,000-year compliance period in previous EPA standards for nuclear waste repositories. They recommended that 10,000 years is not long enough for the compliance period and standards should encompass a significantly longer period. The radioactivity of all of the waste will gradually decline. Some waste elements are very "hot" but have short half-lives. These will release a lot of radioactivity in early years. Other waste elements have much longer half-lives and will continue to generate substantial radioactivity hundreds of thousands of years after disposal. So after the short half-life elements decay, the radioactivity will decline at a slower and slower overall rate. The advantage of Yucca is that it is very far below the surface and yet still very high above the water table. However, the containers will eventually corrode and contaminants could possibly escape through migration in groundwater at that point. The National Academy of Sciences report estimated that the failure would likely occur long after 10,000 years, but at a time when there is still substantial radiation. It concluded that peak radioactive exposures outside the repository would probably occur in several hundred thousand years. Notwithstanding the National Academy of Sciences report, the EPA issued site-specific standards for Yucca limited to 10,000 years. The state of Nevada and environmental groups successfully sued the EPA; the federal appeals court in Washington, DC, held in 2004 that EPA's failure to follow the National Academy of Sciences report recommendation violated the law. The EPA has proposed new regulations with one set of standards for 10,000 years and less stringent standards for 1 million years, but it has not yet adopted new standards. An interesting potential side story about the 10,000-year issue is found at the Oklo site in Gabon, Africa. A remarkable thing about the Oklo reactors is that the highly radioactive waste products from ancient natural nuclear reactions stayed in place without the elaborate containment we use today on nuclear power plant waste. Now, more than a billion years later, everything is contained within a few meters of its source. (For more information, see Alden, n.d.; Cowan, 1976; A prehistoric nuclear reactor, 1973; Smellie, 1995.)

- For now, spent fuel rods from commercial nuclear energy production are being stored at reactor sites in concrete-lined water-filled cooling pools. "Cooled" rods are now being put into safer, dry storage containers. (Defense high-level waste is being stored at DOE defense materials production sites pending shipment to a repository.) Some argue that both the number of sites still storing spent fuel rods in cooling pools and the geographical dispersion of spent nuclear fuel storage generally pose significant safety, monitoring, and terrorism risks. These factors argue for developing an interim storage solution using the best storage technology at a centralized place so that wastes can be more safely stored, effectively monitored, and protected from terrorist attacks. This solution would also remove pressure on the licensing of Yucca. Opponents of interim storage question the advantages of centralized storage, pointing out that it will involve additional expense and transportation of wastes. They also argue that centralized storage will not really be interim and that it will simply remove the incentive to continue to work toward a permanent storage solution. Also, as a practical matter, NIMBY ("not in my backyard") pressures will make it difficult to site an interim storage facility for the most dangerous nuclear wastes. For example, the population near a former uranium mining area near Casper, Wyoming, considered a site, but the state opposed it. The Owl Creek proposal would have provided interim storage for about 40,000 tons. The Goshute Indian tribe in Utah, which claims sovereign state status, has considered a high-level waste storage site but has faced internal opposition as well as opposition from the state government. Some former nuclear weapons sites, notably Hanford (Washington), Savannah River Site (South Carolina), Oak Ridge (Tennessee) and Idaho National Laboratory (Idaho) are logical places for storage sites because they already have waste and a skilled labor force managing it. Surveys (Greenberg et al., 2007b) show that many nearby residents would consider such facilities. But it is likely that there would be adamant state opposition. The NWPA prohibits the DOE from constructing a federal interim storage facility (called monitored retrievable storage, or MRS) until Yucca is operating as a permanent site. The law does not, however, prevent the DOE from looking for and selecting an MRS site, but it has not taken steps to do so. In 1986, the DOE was going to recommend a site near Oak Ridge for an MRS facility, based on a national study of potential sites. Although the local community of Oak Ridge endorsed the recommendation (subject to conditions), the state government strongly opposed it. Two topical reports on the MRS facility were submitted to the NRC, but the DOE did not request an NRC review of the reports.

- Congress imposed a limit on the total amount of waste that can be stored at Yucca (70,000 metric tons of initial heavy metals). Some of this total will be high-level waste from nuclear weapons production but most will be spent fuel rods from civilian reactors. Even if the United States just maintains current levels of nuclear energy production (each plant generates 20–30 metric tons per year), the Yucca repository would fill to this capacity in 2–3 years (according to one of our experts), and there is currently no legal authority or process for taking detailed steps to evaluate a second repository. Yucca's ability to store waste could be extended by the administration's Global Nuclear Energy Partnership (GNEP) proposal for initiating a U.S.–based reprocessing business, which is billed at least in part as a means for reducing the amount of spent nuclear fuel requiring geologic disposal. Another option to extend the operating life of Yucca would be to expand the capacity of the repository so that it can accept more waste than the current statutory limits. Bills have been introduced in Congress in recent years to repeal the 70,000-metric-ton limit, but none has passed.

- A weakness of the original NWPA is that the step-by-step selection process it established was aborted by a political decision to single Nevada out as the nation's repository for the most dangerous nuclear wastes. This action created anger and entrenched opposition to the facility in the state of Nevada, resulting in a great deal of conflict and constant delays in moving forward with Yucca over the past 20 years. However, the delay in licensing Yucca is not entirely unreasonable, because there are many complicated issues to work out and the decision is virtually irreversible. At present, a substantial number of experts, including the national laboratories, have concluded that Yucca is the safest site. But other experts contend that uncertainties remain. The evaluation of Yucca has taken far longer than originally envisaged, but even excessive caution can be appropriate if it leads to the best decision. Some would assert, however, that excessive caution implies gridlock that would prevent the most effective solution from being implemented.
- There has been no comparative evaluation of Yucca's safety relative to other potential sites, a point made by critics who feel that the designation of Yucca as the sole candidate site was wrong. Indeed, any such study is precluded by law. Even if the law is changed, it will be technically and politically very hard to find a second site, if necessary. The volume of waste will eventually catch up and fill Yucca Mountain, particularly if nuclear power generation is expanded in the United States. However, it could be more difficult to significantly expand nuclear energy in the United States until the waste issue is settled. Opponents of nuclear power can always point to the need to deal with the waste. And indeed, some states (e.g., California) have passed laws stipulating that no new nuclear power plants can be established unless and until disposal facilities are in place to receive the wastes they will generate.
- Because Congress was optimistic about how soon a permanent repository could be put into operation, it provided in NWPA that beginning in 1998, the DOE would take over ownership of spent civilian fuel rods for disposal. In return, private utility companies are assessed a fee based on their amount of fuel rod generation. As a result, utilities have paid into the federal government's nuclear waste fund 1 mill (one tenth of 1 cent) per kilowatt-hour sold, but Congress has appropriated only a small proportion of the $27 billion in the fund for the intended purpose of developing a repository.

Because Yucca has yet to open, the federal government has defaulted on its obligation to take over possession of spent fuel rods beginning in 1998. The utilities argued that because the federal government was contractually obligated to take the spent fuel rods, the utilities are entitled to damages and costs related to storing the rods on site. The federal courts have ruled that the DOE does not have to physically take the rods but should pay for economic damages. Over $250 million in payments to nuclear utilities has been ordered thus far in these nuclear waste delay cases. Because of this ongoing litigation, the costs to the federal government of the delay on Yucca and on a permanent storage solution continue to grow. Because of industry claims, the government's liabilities could potentially run to many billions of dollars.

- The main issue with regard to the waste of defense origin is the high-level waste that remains in tanks—mostly underground. In the tanks, many types of waste products were mixed together. Therefore it is hard to characterize the waste. Scientists have to withdraw a sample of the waste, homogenize it, and then assess the level of radioactivity. Fortunately, in terms of potential human exposures, these tanks are found in only a few locations in the United States and are far removed from human populations.
- High-level waste from defense production is usually in the form of liquids or sludge, so it is harder to put it into a more secure form than fuel rods, which are already in a solid form. High-level wastes located at the nuclear weapons site at Savannah River have been vitrified (mixed into molten glass to create a ceramic waste form), and a vitrification facility is being built for the wastes at Hanford. The tanks at Hanford are located close to the Columbia River. Immobilizing wastes in ceramic form is being studied.

What Reporters Need to Know

The core issue is how high-level waste is going to be transported, handled, placed in a safe disposal site suitable for retrieval, and managed safely for a long time. Now that the DOE submitted a construction license application to the NRC in June 2008 for Yucca Mountain, what will be the responses of the NRC, of the EPA, and the states? What will be the position of the strongest environmental groups? How will the courts respond to the inevitable legal challenges? The success of the WIPP effort, in contrast to Yucca Mountain, is an interesting story.

If the Yucca plan fails, then the most logical alternative plan would be interim storage at some of the existing sites. How would an interim storage program be legislated? How will states respond? What will happen if local governments say yes, but their states or adjacent states where waste will pass say no?

Another alternative is the Global Nuclear Energy Partnership (GNEP). GNEP's proposed technologies would consume much of what would otherwise become high-level waste. Some of the uranium and transuranics could become fast reactor fuel (Macfarlane, 2006). How is GNEP being received politically by the Congress? If nuclear power is re-energized in the United States, the amount of spent fuel could increase from an estimated 120,000–130,000 tons for the current plans to a much larger number. A new tunnel could be dug at Yucca Mountain to accommodate more high-level nuclear waste. The George W. Bush administration was considered a friend of nuclear power. How will the new administration react to GNEP and other waste management options? In this context, note that only 1% to 5% of fissionable materials in the nuclear power plant fuel are currently used to generate electricity. Thus, the NWPA requires that the used fuel in the repository must be retrievable during the first 50 years of storage.

Another interesting angle for a story is a comparison of the alternative waste management options being considered by Germany, France, Japan, and other nations. In Germany, the federal government is assuming responsibility for managing and paying for high-level waste management. In the United Kingdom, a government authority has been created to determine how to manage waste and decommission facilities. In the United States, the very small part of the disposal fee paid by utilities has been allocated for the OCRWM to produce a permanent repository. Every nuclear nation is evolving its own waste management policies that are worth investigating and writing about.

Reporters can find good stories in their state's and region's struggles with interstate low-level nuclear waste compacts. Were the failures caused solely by public perception and reaction? Or were there other issues? The few operating commercial low-level radioactive waste sites are able to charge very high disposal rates, which is another potentially interesting story for journalists. That is, how much are we paying per unit of volume or activity to control these wastes? A comparison of nuclear waste low-level disposal costs with other hazardous wastes would be interesting to pursue.

Misunderstandings

Despite a common public perception that nuclear waste is extremely dangerous to human health, in fact, there is very little harm or risk posed by nuclear waste because there are few if any pathways of exposure to the general public. The greatest risk is posed when the waste is moved or transported, and there are potential risks to people living close to waste storage sites. However, the U.S. Department of Transportation and the NRC have established strict safety standards for shipping wastes, and storage of nuclear wastes is tightly controlled. This is not to say that the nuclear fuel cycle is risk free, but a myth has been created that nuclear fuels and wastes are much more dangerous than their alternatives and other substances. Still, many people worry not about properly handled nuclear waste today but about the inability to ensure that natural and human systems will behave as predicted hundreds of thousands of years into the future.

In addition, waste that is labeled as radioactive is not always very radioactive, if at all. Commonly, the low-level waste designation is a precautionary measure. For example, if waste originated from any region of an "active area" within a hospital or industrial facility, it will be classified as low-level waste. This designation can often be given to waste from offices with only a remote possibility of being contaminated with radioactive materials. Such low-level waste typically exhibits no higher radioactivity than one would expect from the same material disposed of in a nonactive area. Some low level waste can contain materials of higher radioactivity or biologically active materials. Therefore, it is required to be monitored and safely handled.

Pitfalls

The United States could expand its nuclear power capacity to generate 30% or more of our electricity needs. (Check the Energy Information Administration [EIA] Web site [*www.eia.doe.gov*] for recent EIA reports, which are frequently updated.) What would be the implications of this nuclear power increase on politics, the economy, the environment, and specifically on nuclear waste management policies? Yucca Mountain is an important issue, but it is far from the only one. Other good stories involve interim storage and current storage at nuclear weapons sites and at commercial nuclear power plants. The biggest pitfall would be to ignore these complex but important issues.

For those interested in reporting more on the Yucca Mountain issue, we recommend starting with MacFarlane and Ewing (2006). The Licensing Support Network (LSN), the U.S. Nuclear Waste Technical Review Board (NWTRB), and the NRC's Advisory Committee on Reactor Safeguards have all prepared documents relevant to Yucca that are available on the Web (see "References and Other Resources" at the end of this brief).

In reporting stories related to nuclear waste policy, it is important to consider the full story, including both sides of the policy issue and the historical context of the problem. For example, if the article concerns nuclear waste disposal, are your sources or facts conveying opposition to disposal at Yucca? If so, you can discuss the alternative with readers, that is, that storage of waste will then have to occur at dispersed sites across the country, most of which, including 104 nuclear power plants located at 64 sites, are much closer to human populations than Yucca.

Finally, the Atomic Heritage Foundation, founded in 2002, is dedicated to the preservation of the legacy of the Atomic Age as context for understanding present scientific, technical, political, social, and ethical issues. Journalists who cover areas near any of the more than 130 sites where nuclear weapons were developed, tested, and manufactured are likely to find interesting human stories to tell about this legacy.

References and Other Resources

Alden, A. (n.d.). *The Oklo natural nuclear reactor. geology.about.com/od/geophysics/a/aaoklo.htm.*

Atomic Heritage Foundation home page. *www.atomicheritage.org.*

Berlin, Robert E., & Stanton, C. C. (1989). *Radioactive waste management.* New York: Wiley.

Burns, M. (Ed.). (1988). *Low-level radiation waste regulation: Science, politics, and fear.* Chelsea, MI: Lewis Publishers.

Cowan, G. A. (1976). A natural fission reactor. *Scientific American,* 235, 36.

Energy Information Administration (EIA) home page. *www.eia.doe.gov.*

Government Accountability Office. (2006, March). *Yucca Mountain: Quality assurance at DOE's planned nuclear waste repository needs increased management attention.* GAO-06-313. *www.gao.gov/new.items/d06313.pdf.*

Greenberg, M., Lowrie, K., Burger, J., Powers, C., Gochfeld, M., & Mayer, H. (2007a). Nuclear waste and public worries: Public perceptions of the United States major nuclear weapons legacy sites. *Human Ecology Review,* 14(1), 1–12.

Greenberg, M., Lowrie, K., Burger, J., Powers, C., Gochfeld, M., & Mayer, H. (2007b). Preferences for alternative risk management policies at the United States major nuclear weapons legacy sites. *Journal of Environmental Planning and Management,* 50(2), 187–209.

Florig, H. K. (2006). An analysis of public interest group positions on radiation protection. *Health Physics,* 91(5), 508–13.

Holt, Mark. (2006). *Civilian Nuclear Waste Disposal.* Congressional Research Service Report for Congress. *www.fas.org/spp/starwars/crs/92-059.htm.*

Institute for Energy and Environmental Research. (2007, October 31). *Comments of Dr. Arjun Makhijani . . . [from] hearing on "Examination of the Licensing Process for the Yucca Mountain Repository." www.ieer.org/comments/waste/yucca071031.pdf.*

Licensing Support Network home page. *www.lsnnet.gov.*

MacFarlane, A. M., & Ewing, R. C. (Eds.). (2006). *Uncertainty underground: Yucca Mountain and the nation's high level nuclear waste.* Cambridge, MA: MIT Press.

Macfarlane, H. (2006). The paradox of nuclear waste. *Radwaste Solutions,* September–October, 32–36.

National Research Council. Nuclear Radiation and Studies Board home page. *dels.nas.edu/nrsb.*

National Safety Council. Environmental Health Center. (1998). *A reporter's guide to the Waste Isolation Pilot Plant (WIPP). www.nsc.org/ehc/guidebks/wippradi.html.*

National Safety Council. Environmental Health Center. (2001). *A reporter's guide to Yucca Mountain. downloads.nsc.org/PDF/yuccapdf.pdf.*

Nuclear Information and Resource Service home page. *www.nirs.org.*

Nuclear Information and Resource Service. Radioactive Waste Project. *Yucca Mountain, Nevada: Proposed high-level radioactive waste dump targeted at Native American lands. www.nirs.org/radwaste/yucca/yuccahome.htm.*

Nuclear Waste Technical Review Board home page. *www.nrtrb.gov.*

A prehistoric nuclear reactor. (1973, January). *Chemistry,* 24.

Reid, H. (2007). *Yucca Mountain* (and related press releases). *reid.senate.gov/issues/yucca.cfm.*

Smellie, J. (1995, March). The fossil nuclear reactors of Oklo, Gabon. *Radwaste* (special series on natural analogs), 21.

U.S. Department of Energy. Office of Civilian Radioactive Waste Management home page. *www.ocrwm.doe.gov.*

U.S. Department of Energy. Office of Environmental Management home page. *www.em.doe.gov.*

U.S. Environmental Protection Agency. Radiation Protection programs home page. *www.epa.gov/radiation/programs.html.*

U.S. Nuclear Regulatory Commission home page. *www.nrc.gov.*

U.S. Nuclear Regulatory Commission. Advisory Committee on Nuclear Safeguards home page. *www.nrc.gov/about-nrc/regulatory/advisory/acrs.html.*

Wilson, P. D. (2002). *The nuclear fuel cycle: From ore to waste.* New York: Oxford University Press.

Monitoring and Surveillance of Nuclear Waste Sites

Written by Karen W. Lowrie, based in part on interviews
with James H. Clarke and Michael T. Ryan, with comments
by Keith Florig

Background

The purpose of an environmental monitoring program at a radioactive waste management or disposal facility is to protect workers, the environment in and around the facility, and the people in it. Monitoring programs are designed to ensure that a disposal facility is in compliance with the U.S. Nuclear Regulatory Commission (NRC) (Licensing Requirements for Land Disposal of Radioactive Waste, 10 CFR pt. 61) and the generally applicable radiation protection standards of Environmental Protection Agency regulations. All waste disposal sites are required to conduct monitoring and surveillance.

Types of Monitoring

Monitoring of nuclear waste sites can be divided into three main types: (1) operational, (2) personal, and (3) environmental. It is fair to say that some type of monitoring is going on all the time. Operational monitoring is designed to assess the basic operations of the site, such as receipt, inspection, offload from the transportation equipment, and disposal of waste materials. It is typically performed with direct reading of instruments and surveys designed to detect even small amounts of radioactive material. Operational monitoring, which begins with pre-operational monitoring to establish a baseline for a site, also includes monitoring associated with reassembling and inspection of transportation equipment prior to its departure to the next waste pick-up location. One reason for these monitoring activities is to ensure that what has been shipped conforms to all waste acceptance requirements of the licenses and permits for the facility. Another is to collect information that is pertinent to the occupational radiation protection program.

Personal monitoring involves the monitoring of workers and other people with possible exposures at the site. The purpose is to ensure that personal radiation exposures are in conformance with all license conditions and regulatory requirements. This usually involves the wearing of radiation exposure recording devices and the use of bioassay (sampling of excreta) and measurements for radioactivity residing in the body (whole body counting) for workers.

Environmental monitoring involves the routine periodic monitoring of the air, soil, vegetation, surface water, and groundwater within, at the boundary of, and in the near environs of the disposal facility. Environmental monitoring identifies and quantifies contaminant releases. It has two purposes. One is to ensure that the environment at the boundary of the facility and beyond is protected and the second is to ensure that all requirements of the facility's operation licenses and permits are met with regard to the radioactive materials. The air, water, or soil is monitored at locations and with frequencies typically approved by the regulator and documented in formal environmental monitoring plans. An adequate monitoring program consists of measuring concentrations of radionuclides and chemically toxic substances in environmental media. In addition, the composition of the buried waste must be known. Soils, crops, and animals (both domestic and game) are sometimes but not always tested for changes in the levels of radioactive materials present in their bodies.

Air and water at and around the site are continually monitored so that if any radioactive materials are released from the disposal facility during operations, they will be detected. Radiation monitors are extremely sensitive and can detect low levels of radioactive materials, making early detection of releases possible and allowing for remedial action to be taken before any release threatens public health and safety.

Hydrological monitoring involves both surface water and groundwater characterization. Surface water is water found above the ground, such as lakes, streams, and rainwater. Groundwater is water found below the surface. For groundwater, data such as the depth of the aquifers and the rate and direction of groundwater flow are gathered. These data are used to determine how quickly radioactive materials introduced into the groundwater move in the groundwater and surface water systems. Surface water data include information needed to minimize the infiltration of water into the radioactive waste site.

Meteorological monitoring is used to determine the atmospheric dispersion of radioactive materials if an airborne release occurs. Some examples of the conditions monitored include wind speed and direction, precipitation, relative humidity, temperature, and barometric pressure. Weather stations are often installed at or near waste disposal facilities. The on-site monitoring of radioactive material in the air can provide an early indication of any releases. Off-site air monitoring is conducted to detect any changes in the background radiation levels so that background values, against which operational data are compared, can be adjusted. It can also indicate the presence of radioactive material from other sources.

Monitoring is carried out during the phase that a site is operational and often continues into the postoperational or stewardship phase, after a site or a portion of a site is closed and capped with an engineered cover above the waste material. The facility's license holder, that is, the company that operates the facility, must continue all monitoring activities in accordance with its licenses and permits. This period may last for 5 years or more after closure. In addition, the site owner or ultimate steward, usually the federal or state government, will continue some monitoring activities for as long as the state deems necessary after closure. After closure, groundwater is the most likely transporter of radioactive materials from the site. As a result, many of the post-operational monitoring activities concentrate on groundwater. Air, soil, and vegetation are monitored as well. The data collected during this phase are compared with similar information collected during the operational and pre-operational monitoring programs. The comparison of data can assist environmental scientists in analyzing trends that may benefit other existing or proposed facilities. Postclosure monitoring programs will be strongly influenced by the operational monitoring results. In some respects, this phase will be easier since the migration pathways should be well known and the number of radionuclides of concern reduced by radioactive decay. Also, monitoring data that has been collected for many years can be analyzed for trends to allow for more accurate predictions of compliance in the future behavior of the facility and better assessments of long-term performance.

Monitoring Methods

Environmental monitoring is done either at the point of release (effluent) or through ambient samples taken at various locations around the site. Effluent monitoring measures contaminants at the point of release. It can include both airborne emissions and liquid discharges. This monitoring is necessary because both site workers and surrounding populations are most likely to be exposed through the air or water pathways, such as inhalation or ingestion. Environmental surveillance measures contaminants that may have spread into the environment. Most sites measure both radiological and nonradiological contaminants. Ambient gamma radiation levels may also be measured on site, at the site boundary, and in surrounding communities. For a very large site, the network can cover very large areas. For example, the Savannah River Site program extends up to 100 miles from the site with tens of thousands of samples collected annually.

Surface samples of soil, sediment, water (seepage basins, streams, and rivers), food products, such as fish, and vegetation are taken directly. Groundwater is extracted from monitoring wells placed in various parts of the underground water system up and down gradient from facilities that could release contaminants. Monitoring and surveillance can be done in ways other than by measuring effluent and taking ambient samples. It is also important to inspect the condition of the engineered or physical controls and barriers to ensure that they are not damaged and are functioning properly. For example, one type of monitoring tool is called the "water balance." This is a measurement of how much water runs off the surface of the site compared with how much infiltrates through the cover. The goal is for no water to infiltrate the cover.

Monitoring Requirements and Programs

The licenses and permits for a facility are the key controlling documents for a facility's operation. These are based on the controlling regulations, the promulgation of which are authorized by controlling laws. (See "Nuclear Waste Policy.") Waste sites are required to have detailed monitoring plans that include how, when, and what is monitored, and how data are recorded and reported. When monitoring data shows levels that exceed recommended minimum levels for a contaminant or reveals that a barrier or control is not functioning properly, corrective action must be taken. The monitoring data provide an early warning to trigger response to unplanned releases. A contingency plan should identify ranges of action levels and the corresponding responses to be taken if unexpected contaminants are detected. Responses could be used to compare monitoring results against a standard, to resample, or to conduct an investigation to determine the source. The key is that facilities monitor to prevent noncompliance. If the facility is open, it is possible that operations will be suspended until the violation is corrected.

Many states have become "agreement states" and have taken on the responsibilities of regulation of radiation, radioactive materials, and radioactive waste disposal. Agreement states develop regulations that are compatible with the NRC regulations. For details on the Agreement States Program, see NRC, n.d. Thirty-four states are agreement states, two have signed letters of intent to become agreement states, and 14 are not agreement states.

Site Security to Reduce Exposures

For additional security, all waste sites have fences and gates around waste storage areas, with access restrictions. Security typically involves fencing with appropriate warning and caution signs. Also, physical security is complemented by security personnel with regular monitoring of security using physical inspections and remote sensing cameras to maintain control over the facility. Site surveillance is typically conducted 24 hours a day 7 days a week. Surveillance is necessary for two reasons: to keep hazardous constituents from getting out; and to keep people from getting in. These measures will include both engineered and institutional barriers. Engineered barriers include the design of the trenches, such as liners and covers, and the form of waste itself. For example, wastes in solid form will not travel as readily as they would in liquid form. Institutional forms of control include deed restrictions, signs, and ownership retention. (See "Long-Term Surveillance and Maintenance.")

What Reporters Need to Know

Most waste sites have been operating for a long time (since the 1970s or 1980s), so there should be a lot of information available from public information offices concerning monitoring records. All monitoring requirements (type, location, and frequency and reporting requirements) are specified in the licenses, permits, and governing regulation with which the operator must comply. Many sites publish an environmental report each year with detailed results from monitoring and surveillance programs. This report should be made available to the news media, as well as to regulators and the general public. Reporters should also check with state authorities to ensure that they are satisfied with the quality of the measurements being reported to them.

Reporters need to distinguish whether the site is an open site or a closed site. Open sites are currently accepting waste and have strict acceptance criteria with required documentation about the type of waste that is entering the site and how it is properly stored. At open sites, waste is handled, so worker health should be monitored as well as the environment. At closed sites, a program of long-term monitoring is required.

During the operational phase of the disposal facility, the concentration of radioactive materials will continue to be monitored and compared with these baseline levels. This comparison will be used to ensure that the disposal facility is retaining the waste. If an increase in radioactive material is found, the point of release can be located and remedial action can begin. So the most important question to ask with regard to monitoring data is: Do the reported measurements indicate that the facility is in conformance with all the applicable regulations and requirements of your licenses, permits and regulations? If not, then you need to follow up by asking: What is in nonconformance? What are the implications to worker health and safety, public health and safety and environmental protection because of the nonconformance? Finally, it is important to ask: What will the site do to bring the situation into conformance with the license, permit, and regulatory requirements?

Pitfalls

When monitoring results are presented, they are often not compared with the level of naturally occurring background radiation, or the baseline-level concentrations. A very small amount of reported contamination may produce public outrage, but the important factor is whether the exposure is potentially harmful or not. People tend to be most comfortable when no level of contaminant is present, even though the environment has naturally occurring radioactive materials in it everywhere. (See "Radionuclides.")

Also, it is important to keep in mind that waste sites are required to monitor for radioactive materials and to inspect the physical condition of site infrastructure, with many standards and required procedures. They cannot operate without this. If you are preparing a story about a waste site, these should not be omitted from your background research.

Another common problem is that terminology is misused and often confuses the story. Radiation is mixed up with radioactive material. The quantities and units used to describe these and other measures of interest are a specialized jargon. Reporters should ask for definitions and clarifications.

References and Other Resources

Conference of Radiation Control Program Directors home page. *www.crcpd.org.*

Health Physics Society. Ask the Expert. *hps.org/publicinformation/asktheexperts.cfm.*

Low-level waste management handbook series: Environmental monitoring for low-level waste disposal sites. (1990, February). DOE/LLW-131G, Vol. 1. Idaho Falls: EG&G Idaho.

McCutcheon, Chuck. (2003). *Nuclear reactions: The politics of opening a radioactive waste disposal site.* Albuquerque: University of New Mexico Press.

Ryan, M. T., Lee, M. P., & Larson, H. J. (2007). *History and framework of commercial low-level radioactive waste management in the United States.* ACNW white paper, NUREG-1853. Prepared by Advisory Committee on Nuclear Waste, U.S. Nuclear Regulatory Commission. Washington, DC: U.S. Government Printing Office.

U.S. Environmental Protection Agency. *Radiation protection, general information. www.epa.gov/radiation.*

U.S. Nuclear Regulatory Commission (NRC). Office of Federal and State Materials and Environmental Management Programs. (n.d.). *State regulations and legislation. nrc-stp.ornl.gov/rulemaking.html.*

U.S. Nuclear Regulatory Commission. Fact sheets on nuclear-related topics. *www.nrc.gov/reading-rm/doc-collections/fact-sheets/.*

(See also state agencies responsible for radioactive waste sites.)

Impact of Radionuclides and Nuclear Waste on Nonhumans and Ecosystems

Written by Karen Lowrie, based in part on an interview with Joanna Burger, with comments by Keith Florig

Background

The prairies, wetlands, deserts, and forests within and surrounding some U.S. Department of Energy (DOE) nuclear weapons facilities are contaminated with radioactive materials released from the facility. Often these lands are so far from population centers that they pose little risk to humans. Radiation exposure to the animals and plants that inhabit the land, however, can be of concern under some conditions. When a radioactive substance is released into or present in an environment, it can affect anything in the environment that is a receptor of that substance. Often news stories will focus on effects of radioactive contaminants or nuclear waste remediation activities on the human workers or human populations nearby, or on the damage to natural resources such as soils or groundwater. However, nonhuman living things (organisms), such as other animal species and plants that are within a pathway of exposure, can also be impacted. An exposure could occur, for example, when the animal ingests, absorbs, or breathes a radionuclide, or when a plant draws a contaminant out of the soil. The effect varies in accordance with the dose, period of exposure, and vulnerability of individual organisms. For almost all contaminants, some organisms are more susceptible than others.

It is important to consider ecological impacts, because ecological resources are very important and prominent at many of the large DOE nuclear weapons sites. The DOE left buffer lands around the production and research facilities at its sites, and many sites were placed in remote locations in the first place because of the need for secrecy and the need to place them away from human populations. Approximately 79% of DOE land is in buffer areas, providing habitat and intact ecosystems, some in areas especially protected for research, preservation, or recreational uses. Thus, the untouched areas in the buffer lands are actually in better shape ecologically than the surrounding areas that have been cleared for agriculture, factories, or houses. And further, since the

DOE left the land alone for more than 50 years, it actually recovered more than the surrounding lands and, in some instances, has the only undisturbed biomes in the region (that is, habitats unique to that region, such as pine barrens or shrub-steppe). The DOE's mission includes environmental restoration, and DOE's stewardship program includes ecosystem management and integration of economic, ecological, social, and cultural factors in land use decisions (DOE Order 430.1).

Before the initiation of land-disturbing or building and structure modifications at the site, archeologists conduct cultural resource surveys or historical evaluations to identify important cultural and historical resources, evaluate the cultural and historical significance, and assess impacts. Native American representatives also conduct cultural assessments of proposed land disturbances to identify resources that may be of religious or cultural significance to American Indians. (For example, see DOE, 2002, for the Yucca Mountain final environmental impact statement [FEIS].)

Ecological risk assessment has been developed to evaluate current or potential damage to ecosystems and their biota (species, populations, and ecological processes). The desired endpoints are ecosystem health, integrity and sustainability, and survival of rare species. To determine risks to ecosystems, scientists use a five-step process. First, they characterize the ecosystem in terms of its components and its functions. Second, they characterize the environmental contaminants for their constituents, toxicity, and form. Third, they assess the exposure (dose or potential dose) to an organism, relative to its body weight. Fourth, because it is impossible to monitor every species, they identify indicator species at different levels in the system (producer, consumer, and decomposer) and biomarkers (substances in body tissue or fluids) that indicate exposure, effect, and susceptibility. For example, a top-level carnivore could be selected as a bioindicator of ecosystem health and a biomarker could be an enzyme that is collected from blood or urine. Fifth, ecological risk assessors need to conduct biomonitoring and surveillance to measure the numbers of organisms, mortality rates, reproductive rates, and biomarkers. Surveillance can also include field observations of behavior or food-web relationships. If monitoring data is reported, to have confidence in the data, it is important to see that QA/QC procedures were followed.

Identifying the Issues

Some of the concerns related to the impact of nuclear materials on nonhumans include exposures to top-level predators through ingestion or through species that are used for medicines, and the health and safety of wildlife populations that may be eaten, hunted, photographed, or enjoyed, or that are just part of the environment. Threatened or endangered species may be of special concern.

Stripping the top layer of soil to reduce radiological contamination can harm existing animal and plant populations more than the radiation levels that existed before remediation. The public could ask whether remediation is going to destroy intact ecosystems that will never recover because of the degree of soil or other disruption. Remediation could be so extensive, for example, that it would destroy the seed bank and allow invasive species to move in, eliminating native species. People might also worry about movement of nuisance and other animals out of the area to escape remediation. When there is remediation, animals that can move will, and they may move into nearby suburban areas.

What Reporters Need to Know

1. *Which organism or organisms are at risk?* It is important to first understand what species are in the ecosystem (the plants and animals that are present in the area). The impact will vary in accordance with the life stage or age of the organism that could potentially be impacted, and will depend on the numbers and spatial distribution of the organism. It is always of interest to know what endangered and threatened species are present (both state and federal), and whether populations on site are critical within the state or region.
2. *What is the level, location, and distribution of each radionuclide?* It is important to find out the level of each radionuclide in the environment both at the time of the release event (if it is an accident) and over a longer period. This information should also include the spatial distribution of the contaminants. Also, which organisms will be exposed depends on whether the contamination is in the soil, groundwater, or sediment running under and along waterways, or in surface water.
3. *What are the likely effects to sensitive individuals?* Some animals are more susceptible than others. So the important question is to determine whether the levels of contamination are sufficient to cause harm to the most sensitive individuals, especially to threatened or endangered species. Radionuclide effects range from sublethal effects on behavior and physiology, to death (either immediate or later). Effects can be chronic or acute.

What Are the Risks?

The amount and the target of harm depends on whether you are most interested in the risk from the particular radionuclide or the risk from the remediation itself. For radionuclides, what part of the ecosystem is likely to be harmed depends on the system (air, water, groundwater, or soil) that was most affected. In general, there is a wide range of effects levels for different radionuclides in different organisms. The reporter will need to first know the radionuclide and the species of interest and then find out the published effects level for that organism (or group of organisms). The most important effect for organisms is the effect on reproductive success and well-being. In most instances, the risk is limited to areas with high levels of radionuclides (such that people might also be affected). With minor levels of radionuclides, the ecosystem might have recovered from the physical damage of building or operating the nuclear facilities.

High-level predators are usually more at risk than other species, but invertebrates are also at risk; thus, each group has to be looked at independently. Aquatic animals are particularly at risk because contaminants move quickly through water. Some animals have higher exposure—such as those that do not roam far from the contaminant source (have a limited range) or spend more time in the actual medium (i.e., fish in water). In terms of ecosystems, more complex systems (more variety of animals and plants) are usually less easily disrupted. Also, even if ingested, contaminants may not all end up in the individual. Whether or not they do depends on intestinal tract and form or condition of substance

In cleanup (remediation) activities, the greatest risk to an ecosystem is from the actual physical disruption that destroys the ecosystems themselves. Often, the removal of soil is more detrimental to species than the levels of radionuclides or other contaminants. Leaving waste sites alone is often best for species that are already coping with the environment as it is.

The risks are going to be decreasing if there is no new contaminant release, because radioactive substances will gradually decay. However, for some that are long-lived (i.e., have long half-lives—see "Radionuclides"), there will be no appreciable decrease for decades or even centuries.

Remediation activities, however, have the potential to re-expose individuals because it disrupts groundwater, water, sediment, and soil, potentially releasing contaminants back into the environment. Also, off-site risks (away from the waste site or release point) can increase if there is an increase in hunting or fishing allowed in the vicinity, if herbs are collected from the area for medicine and cultural or religious practices, or if animal populations increase so that more animals could be moving off site.

For ecosystems, the best approach is to leave them alone, unless there is demonstrable damage to some key species in the ecosystem that needs to be addressed. The next best approach is to do minimal remediation such that there are no new roads built through the system, little disruption of soil, and as little destruction of trees and the ecosystem as possible. If there are high levels of radionuclides in the organisms, there are only two options. One is to kill or otherwise get rid of the organisms in the area and then reintroduce uncontaminated ones back once the system is clean. The other is to clean the environment and let nature take its course (with no new exposure). The result will be the eventual die-out of contaminated individuals.

Pitfalls

Remember that there is a difference between contaminant levels and actual exposure. (For example, even if there are levels of contaminants in deep groundwater, it cannot get to eco-receptors, and thus there is no risk.) There is a difference between exposure and risk. (Radionuclides need to get *into* the body and to be at levels that cause harm.) There is a difference between risk and actual harm. (There might be a risk, but it is so low there is no harm.) There is a difference between harm to an individual animal and ecological damage. That is, organisms could experience some harm, but if it does not affect their *reproductive output* it is not harming populations at the ecosystem level. Humans are the exception: Historically, society invests much more to protect people than to protect nonhuman species. But for ecological receptors, society generally intervenes only when populations are threatened, because the unit of interest is the population and *not* the individual. (Exceptions are endangered and threatened species where each individual is important.)

Remember, also, to consider whether the ecological resources affected are ones that are eaten or otherwise used by people or by top-level predators that roam into the area but are not part of the immediate area. In other words, ecological receptors can move and interact with other ecological receptors, so the effect is not limited to the individual organism that is exposed.

References and Other Resources

Burger, J. (2000). Contaminated Department of Energy facilities and ecosystems: Weighing the ecological risks. *Journal of Toxicology and Environmental Health,* pt. A, 63, 383–95.

Burger, J. (2007). A framework for analysis of contamination on human and ecological receptors at DOE hazardous waste site buffer lands. *Remediation,* 17, 71–96.

Burger, J., Leschine, T., Greenberg, M., Karr, J., Gochfeld, M., & Powers, C. (2003). Shifting priorities at the Department of Energy's bomb factories: Protecting human and ecological health. *Environmental Management,* 31(2), 57–167.

Linthurst, R. A., Bourdeau, P., & Tardiff, R. G. (1995). *Methods to assess the effects of chemicals on ecosystems.* Chichester, UK: Wiley.

National Park Service. *Biology resources. www.nature.nps.gov/biology.*

U.S. Bureau of Land Management home page. *www.blm.gov.*

U.S. Department of Energy (DOE). Office of Civilian Radioactive Waste Management. (2002, February). *Final environmental impact statement for a geologic repository for the disposal of spent nuclear fuel and high-level radioactive waste at Yucca Mountain, Nye County, Nevada.* DOE/EIS-0250. Available from *www.ocrwm.doe.gov/documents/feis_a/index.htm.*

U.S. Fish and Wildlife Service. Division of Environmental Quality home page. *www.fws.gov/contaminants.*

U.S. Forest Service home page. *www.fs.fed.us.*

Whicker, F. W., Hinton, T. G., MacDonell, M. M., Pinder, J. E., & Habegger, L. J. (2004). Avoiding destructive remediation at DOE sites. *Science,* 303, 1615–16.

Long-Term Surveillance and Maintenance at Closed Nuclear Waste Sites

Written by Karen W. Lowrie, based in part on an interview with Raymond M. Plieness, with comments by Keith Florig

Background

The cold war ended nearly 2 decades ago, but the environmental consequences of extensive nuclear weapons research, production, and testing activities carried out in the United States during the cold war will remain with us for many centuries. Although the U.S. Department of Energy (DOE) has spent billions of dollars since 1989 to clean up sites and has succeeded in reducing risks, some level of contamination will remain at many sites where radionuclides cannot be completely removed or are contained in storage. Although some radionuclides, those with short half-lives, naturally decay at a rapid pace into benign substances, some of the long-lived contaminants will be radioactive for tens of thousands of years or more. In addition, other chemical contaminants, such as PCBs, solvents, and heavy metals do not decay appreciably and can remain hazardous in perpetuity.

To ensure that human health is protected for many generations, long-term stewardship must be carried out at sites after remediation is complete. According to estimates made in the late 1990s, more than 100 DOE sites are expected to require some type of stewardship, but the number could change. Some are as small as a football field and others are hundreds of square miles. Some are in close proximity to large human populations and others are in more isolated locations.

Long-term stewardship has been defined as "all activities required to protect human health and the environment from hazards remaining at DOE sites after cleanup is complete" (DOE, 1999). The cleanup process addresses contaminated land, water, groundwater, facilities and materials in accordance with agreed-upon standards and requirements. However, a site could be cleaned up and even closed but still contain known hazards in the form of stored or buried wastes, entombed facilities, such as reactors, or residual contaminants that are left in place, sometimes subject to long-term treatment.

Since 2001, the DOE refers to stewardship activities as "long-term surveillance and maintenance" (LTSM). The two general categories of LTSM are engineered controls and institutional controls. Engineered controls are man-made physical containment systems placed around landfills, vaults, tank farms, or other waste units. They include operating, maintaining, inspecting, and monitoring caps, erosion control systems, environmental sampling, wells, and pump-and-treat groundwater remediation systems. For example, the 62-acre Estes Gulch Disposal Cell near Rifle, Colorado, is designed to hold waste materials from uranium mining and milling operations with a plastic liner and up to 4–9 feet of stone, silt, and clay barrier materials. The Environmental Protection Agency refers to these engineered controls as "active" controls (40 CFR pt. 191), as well as activities such as monitoring the natural attenuation of some residual groundwater contaminants.

Institutional controls are legal or access controls intended to reduce the risk of exposure by ensuring that land- and water-use restrictions are maintained. They include restrictions on land or water use, well-drilling prohibitions, deed notices, easements or other legal advisories or measures, and long-term information management. Access obstacles such as fencing, markers, and signs can be considered passive controls. Federal ownership in perpetuity is itself an institutional control. For example, long-term institutional controls will be necessary in portions of the huge Nevada Test Site north of Las Vegas to prevent inadvertent human exposure to widely dispersed radionuclides and other contaminants resulting from the explosion of thermonuclear weapons.

Long-term support of LTSM activities will be important at both local and national levels to ensure proper and effective implementation. That is, the federal government must continue to evaluate new technologies to improve the effectiveness of remedies or remedial systems (reducing the need and costs for LTSM) and research into better understanding of contaminant movement and impacts of residual contamination. Also, components of LTSM that may need to be addressed and integrated at the local level are emergency response planning, oversight, information management, community land use, and natural resource planning. A recent survey shows that those who live near some of DOE's largest facilities have high expectations regarding stewardship (Greenberg et al., 2007).

LTSM activities are now centralized under the Division of Legacy Management (LM) within DOE. This office handles the DOE's postclosure responsibilities, including long-term surveillance and maintenance, records management, property management, land use planning, and community relations. There are currently 70 sites in 26 states that are officially closed and transferred to LM. Forty of these sites, however, are "records only" sites, meaning that they have been cleaned up to unrestricted use, and LM's main function is to maintain records about the site. The rest of the sites have some type of remaining contamination or a disposal cell that requires some type of ongoing surveillance or monitoring. In some cases, a pump and treat system is in place for groundwater contamination, and in other cases, a set of engineered and institutional controls is in place to protect human and environmental health from any remaining risks. By 2010, it is expected that 98 sites will be under the LM program.

Several existing laws (seven), regulations (nine), and DOE orders (three) require the DOE to conduct monitoring, reporting, record keeping, and long-term surveillance. The specific requirements, design standards, and time frames vary by waste type. LTSM is not unique to the DOE or to radioactive materials. The U.S. Environmental Protection Agency and the Departments of Defense and Interior also practice LTSM activities at sites where some contamination is present, as do many state, local, and private interests that manage facilities with hazardous substances.

Identifying the Issues

One of the most important issues and challenges is building confidence among the various stakeholders at and around the closed sites. Since the hazards and the engineered or institutional controls are likely to be around for centuries, it is important to build trust between communities and the long-term steward, which will be the DOE in most instances (or a successor agency). At many sites, there are people in the community who are leery about trusting the federal government because at some sites there was, and still may be, restricted (classified) information, or past accidents or negative incidents that have created perceptions that the sites will always be contaminated, or that people cannot trust the government to tell the truth about how safe it is. The DOE should continue to work diligently on this issue by communicating often, honestly, and effectively about the conditions on the sites and how local communities can be assured of the safety of the sites. Accountability needs to be ensured with the formation of partnerships between federal, state, and local governments so that all are working together with the same information to maximize the beneficial reuse of sites while maintaining protectiveness of human health.

Future land use is a key factor in determining LTSM needs. If there is no active mission on a site, it could be transferred, meaning that communication with land use authorities is important. As urban development expands toward boundaries, institutional controls must be maintained. This is another issue that underscores the importance of working with local stakeholders to continually improve understanding of local land use conditions, examine likely scenarios, and incorporate these assumptions into LTSM planning.

Another key issue is the stability of continued funding for LTSM. It is essential to maintain necessary levels of funding to ensure maintenance, security, and information management and also important to continue to keep abreast of new information and new applied technologies that can allow site managers to be as proactive as possible to improve site safety. For example, if new science determines that a certain contaminant can be immobilized in the soil before it reaches groundwater, the LM program should have the funds to develop and apply new technology, where applicable, to treat or control this contaminant in the most effective way possible. This important issue necessitates good cost estimates and analysis of different funding mechanisms such as long-term trust funds at the federal level, and attention and awareness to supporting continued funding from local affected communities.

What Reporters Need to Know

A reporter writing a story about LTSM or an LM site should first refer to the LTSM plan for the site. Each plan is site specific and details the contaminants and risks at the site. It is available on the DOE Office of Legacy Management's Web site (*www.lm.doe.gov*). The plan will have the answers to many of the questions in the list that follows. In addition, look at the site annual inspection report, also available at the LM Web site.

- What are the contaminants on the site, where are they located, and what are the risks? What is the likelihood that contaminants will migrate to a water source, particularly if conditions change?
- What remedies and measures are in place to prevent risks? How will the monitoring system detect any failures in the remedy system? What threshold levels must be reached for monitoring to cease?
- What are the roles and responsibilities of the DOE, state, and local agencies with regard to LTSM activities? How will controls be monitored, and for how long? How will LTSM activities be revised to respond to changes in land use?
- How are the LTSM measures adapted to local conditions and communicated through local channels? Are conditions and restrictions on land or water use specified in relevant city, county, and state information systems?

- What is the history of the site in relation to the local community? How can the community get more information and be involved?

What Are the Risks?

The type of risks is directly related to what contaminants are left and what procedures ensure that those contaminants will not reach a receptor. If LTSM activities are sufficient to limit or eliminate any potential exposures to hazards, then the risks should gradually decrease. That is, if the site is remediated to the point of closure, then all of the pathways that would cause unacceptable levels of risk should have been closed, through one or more of the following: removal of the hazard, containment of the hazard (e.g., burial), or access restrictions. As radioactive materials decay, the sites would continue to become less risky. However, if the LTSM activities are not sufficient to protect human health from exposures, then risks could potentially increase. For example, if the hazards are not adequately restricted through fencing, signage, local regulations, or other barriers at a closed site where some contaminants remain, a future nearby resident could use the site for recreational purposes or could drill a well that might cause exposure to contaminants in the soils or groundwater.

The purpose of all LTSM measures is to reduce risk, and they should be designed to work together at any individual site to reduce risk as much as practically, technologically, and economically possible. Any contamination left in place will be isolated in a stable form in a tank or vault or buried or contained within a capped disposal cell, or, in the case of groundwater, it will undergo long-term treatment or monitoring, so that its use will be prohibited until risks are reduced to acceptable levels. The measures to reduce risk are in the form of physical obstructions (caps, sealed vaults), access restrictions such as fencing and signs, and the other institutional controls mentioned earlier. In most instances, there will be several overlapping measures to create layering or redundancy, such that if one mechanism or control fails, another can take its place.

If the stewards responsible for long-term maintenance of these systems are vigilant and responsible in their duties (maintaining barriers, monitoring, communication), then risks will be reduced to levels that are protective of human and environmental health. At some sites, federal ownership in perpetuity is itself an institutional control that will greatly reduce any future exposures, at least as long as the U.S. government is in existence. It is likely that profound changes in human culture and values, land use, science, and technology will occur over vast expanses of time. Also, at some future point, some of the existing engineered or institutional controls at LTSM sites will fail. The hope is that new technologies and developments in knowledge will translate into more robust measures that can be incorporated into LTSM strategies. Transference of information and records to any future stewards or landowners are key parts of long-term stewardship but are difficult to ensure in the present because of the great number of uncertainties. This difficulty underscores the need for continual periodic updating and revision of LTSM planning. LM has adopted the concept of "rolling stewardship," the practice of continually looking 15 to 20 years into the future to assess whether new actions should be taken, and will update its strategic plan every 3–4 years.

Pitfalls

Some groups will charge that LTSM is a means for the federal government to escape its responsibility to fully "clean" sites and shifts burdens to future generations. In reality, some contaminants cannot be removed under current cost and technology constraints. These constraints could change if political priorities shift to support more investment into newer or cheaper technologies that can more effectively and efficiently address remaining contamination.

In addition, news coverage often shows the government in a negative light. When problems arise at a site, there is sometimes not enough discussion about the proactive steps that are taken routinely to protect human health. If a problem occurs, it is important to focus on how all parties are working together on a solution.

References and Other Resources

Carter, L. J. (1987). *Nuclear imperatives and public trust: Dealing with radioactive waste.* Baltimore: Johns Hopkins University Press.

Greenberg, M., Lowrie, K., Burger, J., Powers, C., Gochfeld, M., & Mayer, H. (2007). Nuclear waste and public worries: Public perceptions of the United States major nuclear weapons legacy sites. *Human Ecology Review,* 14(1), 1–12.

U.S. Department of Energy (DOE). (1999, October). *From cleanup to stewardship. lts.apps.em.doe.gov/center/reports/pdf/doc130.pdf.*

U.S. Department of Energy. Office of Legacy Management home page. *www.lm.doe.gov.*

Wiersma, G. B. (Ed.). (2004). *Environmental monitoring.* Boca Raton, FL: CRC Press.

Section 4

Nuclear Weapons, Terrorism, and Nonproliferation

Managing the Nuclear Weapons Legacy

Written by Michael R. Greenberg, based in part on an interview with David S. Kosson, with comments by Richard L. Garwin

Background

Tens of thousands of nuclear weapons have been built, primarily by the United States and the former Soviet Union. The U.S. government began nuclear weapons production under the Manhattan Project and has now moved to the stage of decommissioning many of the weapons and managing the by-products. This brief is about U.S. government efforts to manage its usable nuclear materials from decommissioned and surplus weapons.

Nuclear Weapons

U.S. nuclear weapons are complex devices designed to explode only when authorized, with extensive safeguards to prevent accidental or malicious detonation. There are two basic types of nuclear weapons. An "atomic bomb" uses high-grade conventional explosives and uranium that has been enriched to the extent that it is fissile (a far greater extent of enrichment is required to support a nuclear detonation than is required for nuclear fuel for power production). The high-grade explosives are used to trigger fission of the enriched uranium, that is, the enriched uranium atoms split apart. All U.S. nuclear weapons and many of the others in the world use the artificial element plutonium rather than highly enriched uranium. In either case, the explosives "assemble" a metal shell to become a sphere, or momentarily compress a ball of fissionable material sufficiently to reduce the neutron leakage so that the neutrons released during fission may generate a "chain reaction" that produce more neutrons in each successive generation. (See Walker, 2005, for a well-written account of the creation of the first atomic weapons.)

Current nuclear weapons are "thermonuclear "devices. These weapons use conventional high-grade explosives to trigger fission, which, in turn triggers fusion of heavy hydrogen atoms in tritium gas, that is, atoms are fused together rather than split apart. For context, fusion is what powers our sun. Each of these fusion reactions produces a neutron, and the sudden release of a pulse of neutrons "boosts" the fission reaction to a higher level and multiplies the energy release of the fission bomb by a substantial factor.

A "thermonuclear" weapon uses a fission device or a boosted fission device to prepare a large amount of thermonuclear fuel—usually solid lithium deuteride (LiD) at high compression and temperature to provoke fusion among the heavy hydrogen (deuterium) atoms and between the deuterium and the tritium that is made from the capture of fusion neutrons in lithium. In a thermonuclear weapon, typically half of the energy release comes from fusion and half from fission.

Both forms of nuclear weapons are entirely different from "dirty" bombs and nuclear power plants. A dirty bomb (or radiological dispersal device) uses explosives, like dynamite, to spread radioactive materials. People can be injured or killed by the explosion and debris, and by the spread of radionuclides by a dirty bomb detonation. The dispersal of radionuclides also may cause extensive public fear, even if the levels are low such that there is no immediate health risk. But there is no fission or fusion, and therefore no nuclear blast, and hence the effects of a dirty bomb should be restricted to radiological contamination, though the effects can be damaging to people and property in the local area. (See "Dirty Bombs.")

A nuclear power plant uses enriched uranium (with a lower enrichment level than used for weapons) to create controlled fission that produces heat that is used to turn a turbine, which, in turn, creates electricity. (See "Sustainability.") Nuclear power plants are designed with extensive safeguards so that they cannot explode. They are not bombs but they can have accidents that under worst conditions can liberate much of the radioactivity that has accumulated in the reactor, which would probably be contained in the concrete protective structure. (See "Nuclear Power Plant Safety Systems.")

Managing Surplus Nuclear Weapons Grade Materials

Nuclear power can be part of the solution for managing the nuclear weapons legacy. The tens of thousands of nuclear weapons that are no longer part of the U.S. arsenal must be taken apart so that they no longer can be used as weapons. Each weapon has thousands of components. The vast majority of these are complex electronics. The actual bomb material, the "physics package," has only several hundred components. The fissionable material, consisting of highly enriched uranium or plutonium that was part of the nuclear weapon, is the focus of environmental management. The highly enriched uranium that was in the weapon is diluted with depleted or natural uranium, "down blended," so that it is no longer usable as a nuclear weapons material and can be used as fuel in a conventional nuclear power plant.

The U.S. Department of Energy (DOE) plans to down blend the surplus highly enriched uranium at its enrichment plant in Hartsville, Tennessee (DOE, 2007). Plutonium will be converted in the H-canyon at the Savannah River site and blended with uranium to form mixed oxide fuel. The Savannah River site (SRS) near Aiken, South Carolina, is about 25 miles southwest of Augusta, Georgia. Established in 1950, the 310-square-mile SRS is among DOE's largest facilities, with approximately 16,000 employees and a budget of about \$1.5 billion (DOE, 1996). Its historic mission in the weapons complex was production and separation of plutonium and tritium, but since the end of the cold war, SRS's mission in the DOE complex has been changing to waste management. Down blending is to be accomplished at the H-canyon, which is a 403,000-square-foot facility used for plutonium separation. At SRS, in short, the DOE is attempting to close the circle on the nuclear weapons legacy by using one of its former bomb sites that already has been designed and set aside for nuclear uses to convert the weapons grade material to a fuel.

Some tons of the surplus plutonium will not be turned into reactor fuel but will be treated as waste that will be directed to the vitrification facility at SRS. The 42,000-square-foot, \$2+ billion "vit" plant (the official title is the Defense Waste Processing Facility, or DWPF) blends aqueous nuclear waste with a borosilicate glass compound, which then is poured into a stainless steel container to form massive canisters (they look like logs) that are approximately 8 feet high and weigh about 4,000 pounds. These "logs" are stored on site. The conversion of the waste material from liquid to solid is important because the possibility of exposure through environmental release is greatly reduced when the radionuclides are blended into a melted glass and then cooled to a solid form rather than the liquid or gaseous form. That is, it cannot escape into the air or flow into the ground for thousands or tens of thousands of years or more, by which time the radioactivity will very largely have decayed.

The plutonium surplus from decommissioning nuclear weapons is in a metal form. The metal is planned to be "oxidized," which means adding oxygen atoms to the metal, to change the chemical form from a metal to a metal oxide. Metals commonly oxidize in the air or water, which we know as corrosion in iron. From observation, we have seen that the oxidation of iron takes some time. In contrast, the element phosphorus burns when exposed to oxygen in the air. Plutonium metal can be chemically oxidized in water to plutonium oxide but might first be reacted with hydrogen in a dry process and then converted to oxide.

Oxidized plutonium can be down blended with uranium oxide to form mixed oxide fuel that can be used in a U.S. nuclear power plant (see next section). DOE has begun to test this process with the assistance of France. The United States has shipped plutonium to a French company that processed it and then returned it to the United States for testing. (The French already produce mixed oxide fuel through spent nuclear fuel reprocessing. Reprocessing of commercial reactor fuel in the United States was banned by President Jimmy Carter but permitted again by Ronald President Reagan in 1982. (See PBS, n.d.) Reagan lifted the ban, choosing to "not inhibit it where it does not constitute a proliferation risk." In the United States, the Tennessee Valley Authority (TVA) and Duke Energy are leading the private effort to use reprocessed former military fuel. TVA and Duke are large Southeastern Energy organizations that use a full spectrum of fossil, hydro, and nuclear fuels to provide electrical energy. They operate 13 nuclear power plants in Alabama, North and South Carolina, and Tennessee. In its July 2007 issue, *Site Selection* magazine (Burns, 2007) noted that 30 of the 104 U.S. nuclear power plants are in the Southeast, and 14 of the 28 proposed new applications received by the U.S. Nuclear Regulatory Commission (NRC) were from utilities in the seven southeastern states (Florida, Mississippi, Alabama, Tennessee, Georgia, North Carolina, South Carolina).

Mixed Oxide Fuel

The plutonium oxide from surplus weapon plutonium needs further processing before it can be used as a fuel. Buzz Savage (personal communication) notes that some U.S. utilities, for example, Commonwealth Edison, have used mixed oxide (MOX) fuel in the past on a limited basis to evaluate fuel performance. The DOE's goal is to build a MOX fuel production plant at Savannah River (construction was initiated during summer 2007). MOX is a blend of plutonium dioxide and depleted uranium dioxide. The United States plans to convert 34 metric tons of U.S. surplus weapons grade plutonium into fuel (NRC, 2007). MOX was used initially in 1963 and became a commercial fuel during the 1980s. The French, British, Swiss, and Germans have used MOX fuel in over 30 reactors, and Japan plans to use MOX fuel (Australian Uranium Association, 2007). To date, France, with two operating facilities, is the major producer.

Although the technology has been shown to work, that is, MOX can be created and used in nuclear power plants, the United States is testing the process to demonstrate compatibility with specific nuclear power facilities. The

Catawba nuclear plant near Lake Wylie, South Carolina, owned by Duke Energy, has tested the MOX fuel produced for the DOE by a French company. Groundbreaking for the MOX plant took place at SRS on August 4, 2007. A MOX plant at SRS will centralize the key fuel production technological elements in one location and allow rendering of the most dangerous components of old nuclear weapons legacy no longer weapons usable and available for productive use in nearby nuclear power plants.

Also, the Global Nuclear Energy Partnership (GNEP) is considering building "fast" reactors that could use plutonium as a fuel. (See "Closing the Civilian Nuclear Fuel Cycle.") Specific design choices for these reactors would limit the reactors to use of either a metal or a metal oxide fuel. If the reactor design selected uses metal oxide fuel, then the MOX fuel production plant at SRS potentially may be able to provide start-up fuel for the reactors (this would require further evaluation). If the reactor design selected uses metal fuel, then surplus plutonium may be converted to starter fuel, but this would require using a fuel fabrication facility other than the MOX facility at SRS.

Context for Conversion Effort

The effort sketched in this brief is daunting, but less so than the nuclear weapons legacy. The Brookings Institution (1998) compiled 50 facts about U.S. nuclear weapons primarily from published U.S. government sources. The United States built more than 70,000 nuclear warheads and bombs, 67,500 missiles, 4,680 bombers, and 18 ballistic missile submarines to carry these weapons. Fissile materials produced were 104 metric tons of plutonium and 994 metric tons of enriched uranium. In 1966, which was the peak year, the nation had 32,193 nuclear weapons and bombs. In 1998, Stephen Schwartz estimated that the U.S. government had spent about $5.5 trillion on its nuclear arsenal. When updated for the changing value of the dollar and even allowing for substantially reduced expenditures since the end of the cold war, the historical cost is estimated as over $100 billion a year ($100 billion a year from 1943 to 1998 in 1998 dollars; $124 billion a year on average in 2007 dollars).

Along with this historical context, the immediate driver of this effort is that in July 2007, President George W. Bush set a goal of cutting the number of deployed strategic nuclear weapons in half by 2012. This means that the United States would have between 1,700 and 2,200 operationally deployed weapons by 2012, the lowest number since 1950, which is the range required under the bilateral agreement between the United States and the former Soviet Union. Both countries have agreed to reduce their weapons-usable plutonium reserves by 34 metric tons (WNA, 2008).

What Reporters Need to Know

Reporters need to understand the basics of radioactivity, nuclear fuel sources, nuclear reactors, and the GNEP effort to organize a partnership among nations interested in cooperating in nuclear power technology and recovering spent nuclear fuel (see separate briefs devoted to each topic). Then it is imperative that they recognize the differences among nuclear weapons, dirty bombs, and nuclear power plants.

In addition to the technological challenge, it is important to understand the legal and political contexts, especially involvement of multiple federal departments and groups within each department (see "Nuclear Waste Policy"). DOE has the primary responsibility. Within DOE, the National Nuclear Security Administration is responsible for the weapons and decommissioning the weapons, and Environmental Management focuses on the waste products. DOE's Office of Nuclear Weapons Management, the Office of Defense Programs (DP-22), is responsible for shipments.

Pitfalls

It is a serious problem to confuse nuclear weapons, nuclear power plants, and nuclear waste. Bombs and power plants are very different.

Insufficient attention is paid to the full life cycle of nuclear materials used in weapons and produced as a result of decommissioning. For example, surveys show public concern about transporting nuclear materials. More effort needs to be made on reporting the actual record of transporting these materials. A neglected story is the redundancy of safety systems built into facilities that manage and transport nuclear materials, including the waste products. Some of that information cannot be disclosed for security reasons, but other parts would be good stories for locales with nearby nuclear plants.

Another evolving story is the international effort highlighted by GNEP, which could lead to unprecedented levels of cooperation among the world's nuclear nations. Also, as noted earlier this effort is part of an agreement by the United States and the former Soviet Union to reduce their stockpiles of plutonium. Will the United States–Russia cooperation continue? Zagoroli (2006) reports that Russian government has been reluctant to agree to limiting U.S. corporate liability and costs related to the Russian MOX plant. We should expect continuing and evolving international negotiations to influence the relative success of the effort to convert weapons grade materials to fuel. This will be an ongoing story, with high stakes for the United States because of its massive nuclear weapons legacy.

Within the United States, Congress has assigned responsibility for designing, building, and operating the MOX plant to Shaw Areva MOX Services (formerly DCS, a conglomerate of Duke COGEMA Stone & Webster). But while DCS is a contractor for the DOE, the NRC is responsible for licensing, oversight regarding public health and safety, security, and environmental protection (NRC, 2007). The interactions among these private and public organizations will be an interesting story for the media to follow. Finally, if the effort to redirect military weapons grade materials to fuel fails or moves more slowly than anticipated, the issue may be, as it often is, cost. Zagaroli (2006) reports that the cost of building the MOX plant at SRS has risen from about $1 billion to over $3.5 billion, and it will not be completed until 2015, 6 years later than anticipated. The story here is the high cost of building technologically complex projects, a worldwide problem that merits attention.

A case in point is the MOX plant at the Sellafield nuclear complex in England that was opened in 2002 at a cost of GBP (Great Britain pounds) 470 M (about $940 M at 2008 exchange rate of $2/GBP) and intended to produce 120 tons of MOX fuel annually. The British energy minister, Malcolm Wicks, admits that the plant has produced no more than 2.6 tons of fuel in any one 12-month period (Macalister, 2008).

References and Other Resources

Australian Uranium Association. (2007, May). *Mixed oxide fuel (MOX).* Nuclear briefing paper 42. *world-nuclear.org/info/inf29.html.*

The Brookings Institution. (1998). *50 facts about U.S. nuclear weapons. www.brookings.edu/projects/archive/nucweapons/50.aspx.*

Burns, A. (2007, July). Power punch. *Site Selection. www.siteselection.com.*

Duke Power home page. *www.dukepower.com.*

Federation of American Scientists. Garwin Archive. *www.fas.org/RLG.*

Macalister, T. (2008, March 3). Minister admits nuclear fuel plant produces almost nothing. *The Guardian. www.guardian.co.uk/world/2008/mar/03/nuclear.energy.*

National Nuclear Safety Administration home page. *www.nnsa.doe.gov.*

Public Broadcasting Service (PBS). (n.d.). *Presidential actions: A brief history.* Frontline series. *www.pbs.org/wgbh/pages/frontline/shows/reaction/readings/rossin1.html.*

Schwartz, S. (1998). *Atomic audit: The costs and consequences of U.S. nuclear weapons since 1940.* Washington, DC: The Brookings Institution.

Tennessee Valley Authority home page. *www.tva.gov.*

Union of Concerned Scientists home page. *www.ucsusa.org.*

U.S. Department of Energy home page. *www.energy.gov.*

U.S. Department of Energy (DOE). (1996). *Closing the circle on the splitting of the atom.* Washington, DC: U.S. Department of Energy.

U.S. Department of Energy (DOE). (2007, July 24). U.S. nuclear weapons strategy delivered to Congress. *www.energy.gov/news/5275.htm.*

U.S. Nuclear Regulatory Commission home page. *www.nrc.gov.*

U.S. Nuclear Regulatory Commission (NRC). (2007). *Frequently asked questions about mixed oxide fuel. www.nrc.gov/materials/fuel-cycle-fac/mox/faq.html.*

Walker, S. (2005). *Shockwave: Countdown to Hiroshima.* New York: Harper Perennial.

World Nuclear Association (WNA). (2008, March). *Military warheads as a source of nuclear fuel. www.world-nuclear.org/info/inf13.html.*

Zagaroli, L. (2006, March 8). Program to turn plutonium bombs into fuel hits snags. *www.shns.com/shns/g_index2.cfm?action=detail&pk=NUCLEAR-03-08-06.*

Dirty Bombs (Radiological Dispersal Devices)

Written by Michael R. Greenberg, based in part on an interview
with Detlof Von Winterfeldt, with comments by Richard L. Garwin

Background

A dirty bomb, more formally labeled radiological dispersal device (RDD), is a mixture of conventional explosives and radioactive materials derived from medical equipment, food and blood irradiators, other industrial uses, or nuclear civilian and military waste. When a dirty bomb is detonated, injuries and perhaps deaths will result from the blast and shrapnel. Some of the radioactive material will fall near the detonation site, and the remainder will disperse in the air as a cloud and settle far away from the detonation site. The impact of the nuclear material on the surrounding population and neighborhood will be very site and weather specific. No nuclear fission or fusion takes place.

Identifying the Issues

Dirty bombs have become an issue because they are likely weapons for terrorists. A dirty bomb does not require the technical expertise or funds required to obtain, build, transport, and use atomic or hydrogen bombs. Dirty bombs are a variation of weapons that terrorists already know how to make and use. Dirty bombs can cause health effects for those who are near the blast or who rush to the site and are not protected from exposure to radioactive materials. They are likely to have psychological effects and contaminate large enough areas to cause economic dislocation and stigma.

What Reporters Need to Know

For a pre-event background story, a reporter should probe with two risk-related questions.

1. What sources of radioactive materials are located in the area? How dangerous are the radioactive substances? Are the radionuclides short- or long-lived?

The U.S. General Accounting Office reported in 2003 that there were about 2 million licensed sealed sources of radioactive materials in the United States, and the European Union reported another 500,000 in Europe. Calibration equipment contains low levels (<100 curies, or Ci) of cobalt-60, cesium-137, and americium-241. Medical and industrial facilities use radioactive materials for diagnosis, sterilization, and research (1,000 to 5,000 Ci). Larger amounts of radioactive cobalt-60 (2,500 to 25,000 Ci) and lesser amounts of cesium-137 (50 to 15,500 Ci) are found at blood irradiation facilities. Medical facilities use single and multi-beam radiotherapy. The radioisotopes can be very active (500 to 27,000 Ci of cobalt 60 and cesium 137). Industrial irradiators used to sterilize food for preservation have radioactivity levels of a million or more Ci. Facilities that store or reprocess radioactive materials from nuclear power plants or from nuclear bomb production processes also have nuclear weapon waste with millions of Ci, as well as multiple types of radioisotopes. It is important that these differences in radioactivity be ascertained, as well as differences in half-life and type of radiation. (See "Radionuclides.")

2. How effectively are radioactive materials contained? In a solid form? Enclosed in a tank or encapsulated in a solid form that would limit its mobility? In a liquid form, in tanks below or above ground? In a gaseous form? What is the security protocol for these radiological materials?

In the United States and many other countries, the purchase, use, and disposal of radioactive materials are regulated. Yet there always is some chance of theft and misuse. For example, we are concerned about some medical or industrial irradiation equipment, which although containing lower quantities of radionuclides, are housed in portable devices that could be stolen.

If there is an event or the reporter is trying to estimate the impact of an event, the following questions are important.

3. How many people are nearby who can be potentially exposed?
4. How much exposure is likely and what is the risk to exposed people?
5. What will be the response of local, state, and federal officials? Is it better to stay indoors or to evacuate? Can people, animals, and materials be relocated so that they are not exposed during a radiation release? Where will exposed and injured people and animals be taken? How will contaminated

structures, property, and soil be managed? Who will be responsible? What can an individual do?

6. How will the area be secured until it is safe? How long will the public be kept out of the area? Will access be limited in perpetuity?
7. How will government and responsible parties follow up so that the chances of subsequent exposures are reduced?

What Are the Risks?

The explosion is the first risk. It could kill or severely injure nearby workers (e.g., dockworkers, truckers, security personnel), and people who unfortunately are proximate. If it is a dirty bomb, emergency personnel, such as fire, police, and ambulance workers could be exposed, depending on the rem level. A rem (roentgen equivalent man) is a unit used to measure the radiation dose received by people in a given interval. On average, people receive 300 mrem (millirem) per year, or about 1/3 of a rem per year. A single whole-body CT scan exposes people to more than 1 rem at a time. The exposure of radiation workers is limited to 5 rem per year. Some environmental standards require limiting public exposure to less than 15 mrem per year. Exposures in the millirem range are not likely to produce any health effects at all. Exposures in the higher rem range may produce health effects over the period of many years. Acute effects like radiation sickness and burns occur well above exposure of 300 rem.

The next level of risk results from a plume that could spread over as much as 50 square miles in case of a very large release. Even in this instance, the outer edge of the plume will have exposures of only about 100 mrem for a 4-day dose, and about the same amount for a 1-year dose from the residual radioactivity left on the ground after the plume has passed (the "ground shine"). The impact of the plume or of the ground shine on public health may not be detectable, even after many decades, because 100 mrem is a small increase over the normal exposure of an average U.S. resident of about 300 millirem per year.

The most likely risk is economic stigma resulting from ground shine. There will be evacuations and business relocation, and there will be residual hotspots. Simulation studies done of Los Angeles show that if we try to remove contaminated hotspots to below 15 millirem per year, which is the Environmental Protection Agency standard, the cost could be hundreds of billions and possibly 1 or 2 trillions of dollars. Cleaning up to 100 millirem would still cost billions of dollars, but this simulation illustrates that cleanup standards have enormous consequences. Ultimately, the challenge is costs versus benefit of cleanup. If multiple dirty bomb attacks occur, there will be serious economic repercussions.

The likely targets are valuable economic areas such as ports, tourist attractions, financial centers, and transport nodes. But as we learned during the World War II bombing of Dresden and Tokyo, attacks can be directed at workers and housing to demoralize the public. Overall, there will be some public health repercussions, but the larger risk is to those who own property and operate businesses in the affected areas.

We have learned that terrorists like to do what they know how to do best. Hence, a dirty bomb loaded in a truck is a likely method of attack. A bomb likely would be built by one group, transported to the site, and detonated remotely by another. However, we cannot discount the use of trains, ships, planes, or helicopters in such an attack. Nor, in the case of small amounts of medical radionuclides, can we discount suicide bombers with backpacks or suitcases. In June 2002, the United States arrested José Padilla for his involvement with Al Qaeda planning a dirty bomb attack in the United States. In January 2003, British officials found documents in Afghanistan indicating that Al Qaeda had constructed a dirty bomb and had plans about how to use it.

Dirty bombs scare people and are likely to reduce property values in areas with residual radioactivity. So it might be the choice of terrorists, especially because it is much easier to acquire radioactive materials from medical and industrial uses than to build a nuclear device. The risks of a dirty bomb attack, we believe are increasing. The materials are relatively easy to obtain and the bombs are not notably difficult to build for those who already have experience in building bombs.

The International Atomic Energy Agency (IAEA) points out that almost every nation has radioactive materials in equipment and devices. Some has been stolen, lost, or abandoned. Controlling sources of radioactive materials is critical. This means monitoring and protecting a wide range of materials from less active radionuclides used in hospitals to nuclear waste materials found at storage and reprocessing facilities. The United States has become increasingly engaged in source control. Nuclear processing plants in Russia, Ukraine, India, Pakistan, and now North Korea and soon perhaps Iran have become major concerns.

In this regard, the United States faces a major diplomatic challenge in working with other countries to prevent radioactive materials from being transported on ships and other international carriers. It needs to convince other nations that it is to their benefit to restrict the flow of nuclear materials.

This doubtless means technological and financial assistance as well as diplomacy. (See "International Agencies and Policy.")

The U.S. government has expanded its efforts to intercept radiological materials at ports of entry. Coast Guard officers board merchant vessels and use detectors to look for radiological materials. These efforts have expanded in commercial ports, and new detection equipment is better able to detect efforts to hide radioactive materials. Of increasing concern are pleasure boat docks in the United States and in Mexico, Colombia, and other nearby locations. Inspecting pleasure boat marinas is a major challenge requiring substantial resources.

Using legal measures enhanced since the terrorist attacks of 9/11, the United States has been clamping down on entrance of foreigners. After the Iraq war ends, a trained cadre of bomb-making experts will unfortunately be available to focus on other targets, including the United States.

With regard to stopping or mitigating the impacts of a dirty bomb attack, an attack is likely to occur in phases and, hence, can potentially be discovered and stopped. Someone will need to obtain a job or gain access at the selected source of radioactive materials. Then the radioactive material would be stolen or purchased illegally. The material must then be transported to a warehouse or other place large enough to build the bomb. Simultaneously, someone must study the selected target or targets. Then the weapon must be transported to the target. From the time the radioactive material is obtained to the time of detonation, the radioactive material can be detected.

A dirty bomb plot can also be detected by following evidence regarding conventional bomb-making material, and by preventing the movement of those who build, move, and detonate the weapon. The implications of such efforts, however, are likely to mean substantial resources devoted to monitoring and surveillance of hazardous materials, and people who are likely to abuse them. This, of course, has potential implications for civil liberties.

Misunderstandings

Some reporters could portray a dirty bomb as a public health disaster, as if it were a major accident at a nuclear power facility or a nuclear weapons detonation. Research in Los Angeles shows that we could expect a radiation cloud of 100 millirem up to several miles from the source for 3 to 4 days and some residual ground contamination. This, in itself, would not be a public health disaster. But it certainly would be an economic problem. Although a dirty bomb is newsworthy, overstating the impact would potentially scare people into evacuating when they might be safer staying. The more time they spend outside, the more the dose and the more likely they are to become involved in an accident while fleeing the area.

As much as possible, reporters need to put aside a tendency toward the dramatic in these kinds of situations and focus on details of the situation at hand. For example, remarkably different impacts can result from a bomb with the same hazard potential. On one hand, it can have no radiological-related effects if the bomb was poorly constructed so that there was no release or if it was detonated in an area where there were few people or if the wind blew the blast materials away from people. On the other hand, if it was competently produced, the bomb could release 20% or more of the radiological materials toward a populated area. Misunderstandings can be avoided by obtaining accurate information from trustworthy sources.

Pitfalls

Coverage of this issue today in the United States has been satisfactory. However, we worry about what would happen in an actual event. A major concern is the presentation and visual display of information without the proper context. The public is unlikely to understand the public health risk associated with a 100- rem exposure, or a 100-millirem exposure per year. If they see a map showing exposures in a given area, they are likely to assume that everyone and everything in the plume is at high risk. Anything inside of the plume, even if the risk is minimal, could suffer a long-term stigma. People and businesses might leave and not return. Property values will decline and could remain depressed.

The tendency for journalists will be to want to display any traces of a cloud, even if the risk is minimal. The National Radiological Atmospheric Center in Livermore does real-time simulations. They can quickly provide detailed radiological exposure maps (see *llnl.gov* for a list of papers or call 925-422-4950 for telephone access). But what should be displayed? Should it be the area of 100 rem, which may result in radiation sickness? Or 100 mrem, which is not likely to lead to significant or even detectable health effects? Or 15 mrem, which is detectable by today's instrumentation but will certainly not produce health effects?

If a bomb is successfully detonated, anywhere from 1% to 80% of the radionuclides could become airborne and respirable. The plume created after an event will depend on the specific meteorological conditions at that time in that location. We use mathematical models to predict what might happen. For example, simulations done in Los Angeles for a very large dirty bomb source suggest that a 4-day dose of 1,000 millirem (1 rem) would occur in the area immediately around the detonations and 100 millirem in the larger area under the plume. The ground contamination will increase the exposure if much of it becomes airborne again and is inhaled or becomes absorbed into food or water supplies. Decontamination can reduce exposure from this source. For context, public background radiation averages about 300 millirem a year; a CT scan delivers a dose of 1.3 rem, and the worker radiation standard is 5 rem per year.

Journalists can avoid serious economic and political damage to communities by understanding the implications of exposure to low levels of radioactivity from a dirty bomb. This means that journalists need to work closely with local and state health officials to precisely describe the implications of radiological fallout from dirty bombs.

Resources

Carnegie Mellon University. Center for Risk Perception and Communication home page. *sds.hss.cmu.edu/risk.*

Electric Power Research Institute (EPRI) home page. *www.epri.com.*

International Atomic Energy Agency home page. *www.iaea.org.*

Kelly, H. A. (2002, March 6). Testimony of Dr. Henry Kelly, President, Federation of American Scientists, before the Terrorist Nuclear Threat. Senate Committee on Foreign Relations. *www.tinyurl.com/ywu59a.*

National Academy of Sciences. Nuclear and Radiation Studies Board home page. *dels.nas.edu/nrsb.*

Union of Concerned Scientists home page. *www.ucsusa.org.*

United Nations Scientific Committee on Effects of Atomic Radiation (UNSCEAR) home page. *www.unscear.org.*

University of Southern California. Center for Risk and Economic Analysis of Terrorism Events (CREATE) home page. *www.usc.edu/dept/create.*

U.S. General Accounting Office. (2003). Non-proliferation: U.S. international assistance efforts to control sealed radioactive sources needs strengthening. GAO-093-638.

U.S. Nuclear Regulatory Commission home page. *www.nrc.gov.*

(See also state government environmental or public health agencies, radiation control sections.)

Nuclear Nonproliferation

Written by Michael R. Greenberg, based in part
on an interview with Victor Sidel, with comments
by Richard L. Garwin

Background

Arms-control agreements attempt to manage the numbers, types, development, deployment, and employment of weapons and armed forces. These agreements have failed to prevent the development and deployment of weapons of ever-increasing destructive capacity, although they have slowed their development and proliferation.

A key part of the effort has focused on controlling highly enriched uranium, weapons grade uranium and plutonium, and methods to use them as weapons. Over 60 years ago, the United States, fearful of German intentions to build and use nuclear weapons, was the first to develop, and the only country to use, nuclear weapons. After World War II, Presidents Eisenhower and Truman tried to promote the peaceful uses of radionuclides. Having failed to refocus attention on peaceful use of radionuclides and reach a settlement about nuclear weapons with the Soviet Union, the United States unsuccessfully attempted to guard nuclear weapon secrets, and during the Korean War, the United States seriously considered using nuclear weapons.

For more than 40 years, the world lived uncomfortably with the concept of mutually ensured annihilation, that is, the United States and Soviet Union built and made ready for use tens of thousands of nuclear weapons to deter each other from using them. Thousands of nuclear weapons remain in the arsenals of the United States and Russia and a much smaller number are held by the United Kingdom, France, China, Pakistan, India, and probably Israel and North Korea. The nuclear weapons of the early 21st century are thousands of times more destructive than those used during World War II. The stakes are high and continuing to increase, in view of the potential for use of nuclear weapons by nonstate groups and irresponsible states.

Identifying the Issues

Antiproliferation agreements are our first and best hope for limiting the spread of nuclear weapons and ultimately of eliminating these weapons. Nuclear weapons are capable of causing almost indescribable destruction of people, places, and ecosystems. The blast by itself, the firestorm, burns, debris, and ionizing radiation can kill millions. Fallout can injure and kill people located far from the blast. Managing the impacts of a nuclear weapon exchange or a release of nuclear materials doubtless will necessitate human rights violations. The core issue of this brief is what is being done to reduce horizontal and vertical proliferation of nuclear weapons.

What Reporters Need to Know

Reporters should focus on three topics. First, what is the nation's proliferation stance? In this regard, journalists should distinguish between horizontal and vertical proliferation. The Nuclear Nonproliferation Treaty (NPT) of 1968 committed nuclear and non-nuclear states to a two pronged reduction of risk. The non-nuclear states signed the treaty to refrain from acquiring nuclear weapons, in other words, no horizontal proliferation. In turn, the United States, China, France, the United Kingdom, and the Soviet Union—the original five nuclear nations—agreed to end the nuclear arms race and negotiate nuclear disarmament. The signatory non-nuclear states were promised assistance to develop peaceful uses of radionuclides and also nuclear power.

Although the original five nuclear states have nearly all the weapons, the entry of India, North Korea, Pakistan, and probably Israel into the nuclear weapons group of nations demonstrates that the NPT has not been entirely successful. With regard to proliferation, before reviewing the types of proliferation, we note a distinction between signers and nonsigners of the international nonproliferation treaty. India, Israel, and Pakistan, for example, did not sign. Iran and North Korea did, and consequently the latter received technical assistance from the International Atomic Energy Agency. This poses a clear challenge to the credibility of international treaties and governance. Horizontal proliferation worries people, who believe that diffusion of weapons inevitably increases the likelihood of theft and smuggling by terrorist and rogue nations and increases the chances that regional disagreements could lead to a devastating and uncontrollable nuclear exchange.

Vertical proliferation means that nuclear nations like the United States build more or "better" bombs and delivery systems, as well as missile defenses to protect their countries. For example, the United States and the Soviet Union built multiple independently targeted reentry vehicles (MIRVs) that would carry many more destructive weapons on a single missile, thereby allowing one missile to target multiple locations or hit the same site multiple times. Under President Ronald Reagan, the United States initiated the Strategic Defense Initiative (SDI), known as "Star Wars," which was going to build a defensive shield of interceptor missiles around the United States.

A more recent example of vertical integration is the threat to use nuclear "bunker busters" (B61–1), which supposedly can penetrate underground bunkers. The immediate issue is a threat to use these missiles to destroy Iran's nuclear capacity. These threats are seen as a direct threat to non-nuclear nations that could lead some of them to consider acquiring nuclear weapons for self-protection. Furthermore, a National Academy of Sciences (2005) report emphasizes that tactical nuclear weapons might not penetrate far into the ground and could spread radiation over wide areas.

While small nations would not be able to acquire hundreds of nuclear weapons, we know that a small amount of fissionable nuclear material as part of a bomb can have a devastating effect on public health, the environment, and the economy.

In an 2006 editorial, former Secretary of State Henry Kissinger opposed the strategy of preemptive strikes, which the Bush administration has indicated it might use. Kissinger said that chaos would result if each nation is able to set its own rules for a preemptive strike.

A major problem is what nation or nations are going to take the lead on nuclear disarmament. An attachment to the fiscal year 1994 defense authorization bill prohibits the Department of Energy's nuclear laboratories from developing a precision nuclear weapon of less than 5 kilotons (Hiroshima was estimated at 15 kilotons; 1 kiloton equals 1,000 tons of TNT) because it blurs the distinction between nuclear and conventional weapons (Nelson, 2001). In 2005, Congress blocked further research on tactical nuclear weapons. But stockpiles of these weapons already exist. Overall, horizontal and vertical proliferations are complementary. A failure in one increases the likelihood of a failure in the other. In short, journalists must be familiar with the political positions of their leaders and understand the reasons and implications of those positions.

A second topic of focus for reporters, in the event of a nuclear attack, is to ask about the hazards and risks involved. What kind of weapon or release has occurred? Was it fissionable material or not? Was it above ground? Below ground? What type of radiation is emitted? What is its yield? Is there a plume? How far will it spread? How many people were in the area of the blast? In the area covered by the plume? What kinds of businesses were in the area? Have people been evacuated? How many are left? How are people being treated?

Third, in response to the first two questions, What is the level of preparedness of the state and local governments to respond to an event? During the 1950s and 1960s, preparedness was inadequate. Experts contend that the drills conducted by school children and family would have been ineffective. What plans do state and local governments have and have they been evaluated?

What Are the Risks?

On one hand, as evidenced by recent events in Iran, Pakistan, North Korea, and India, proliferation has occurred as have smuggling, theft, and profiteering or "first-tier" proliferation, primarily from China and Russia but also in the past from France and the United States. The United States and other nuclear nations are also worried about "latent' proliferation, which refers to the concern that nations claim to adhere to their obligations under NPT while secretly developing and acquiring nuclear weapons and delivery systems. The United States is also worried about "second-tier" proliferation, which refers to the concern that nuclear states like China, North Korea, and Pakistan are helping each other or additional states such as Iran or Libya by providing nuclear materials, technology, and weapons delivery systems.

On the other hand, non-nuclear states argue that they have been deceived by their nuclear counterparts. As this brief was being prepared in May 2007, Anwar Othman Albarout, head of the United Arab Emirates delegation, argued that the gap between the nuclear and non-nuclear states has been increasing because the nuclear states insist on vertical proliferation. These important international differences in viewpoint are unlikely to change anytime soon.

Yet, there is an undercurrent of international support based on the awful consequences of the explosion of a bomb. If there is not, there should be. The *Bulletin of the Atomic Scientists* devised a "doomsday clock" in 1947 to highlight the risk. The original setting was 7 minutes to midnight. In 1953, it was 2 minutes to midnight because the Soviet Union continued to test more and more nuclear devices. Now it stands at 5 minutes to midnight. Perhaps the clock exaggerates the risk. But many do not believe so.

There is a palpable tension across the globe in every nation about the wisdom of acquiring nuclear weapons. Some believe that they need weapons to protect themselves, and not just against nuclear weapons, for example, North Korea, Pakistan, and India. Yet other nations with the capacity to build nuclear weapons have not done so or have discontinued their efforts, for example, Sweden and South Africa.

Efforts are both national and international. Security and maintenance of nuclear weapons and fissionable materials varies enormously by country. U.S. facilities are well guarded, apparently with exceptions—six advanced cruise missiles were accidentally flown from Minot Air Force Base to Barksdale, Louisiana, on August 29, 2007, and then sat there on the field, unguarded. Russia has many more security problems, especially in regard to plutonium and weapons grade uranium. There is worldwide concern about proliferation from and to China, Pakistan, India, North Korea, and Iran.

The International Atomic Energy Agency has the capacity to find weapon usable materials. However, nations have to be willing to accept international inspectors, which as we know from recent history in Iraq and in North Korea, they may not. The NPT, as described earlier, is a key international institution. The 1996 Advisory Opinion of the International Court of Justice, many decades of UN General Assembly resolutions, and the Proliferation Security Initiative all call for progress on nuclear disarmament. But these are advisory and nonbinding. Some success has been realized in bilateral agreements between the United States and Russia. The Strategic Arms Limitation Talks (SALT) and Strategic Arms Reduction Treaty (START) have reduced the number of active nuclear weapons held by the United States and the Soviet Union and its allies, from about 70,000 to less than half that number. Still fewer, estimated about 2,000, are ready to be launched in minutes by the United States and Russia.

While our focus has been on the proliferation of nuclear weapons, there is risk associated with radionuclides, in general. Absent adequate controls and security, a variety of activities can release radionuclides into the environment. An airplane could be crashed into a nuclear power plant, a nuclear waste management facility, and locations where spent fuel assemblies are stored. While an attack could be launched on such facilities, these would not be easy attacks to accomplish. However, the consequences could be substantial, including direct exposure of workers and of people living nearby, and some radioactive material carried aloft. Dirty bombs would seem to be the easier to construct and use. (See "Radionuclides" and "Dirty Bombs.") Securing and guarding radionuclides, especially fissionable materials, is an enormous challenge for this generation.

Misunderstandings

Reporters need to understand the difference between fissionable nuclear materials and other nuclear materials. As defined in a 1995 DOE *Safeguards and Security Glossary of Terms,* nuclear materials are:

a. All materials so designated by the Secretary of Energy. At present, these materials are depleted uranium, enriched uranium, americium-241, americium-243, curium, berkelium, californium-252, plutonium 238–242, lithium-6, uranium-233, normal uranium, neptunium-237, deuterium, tritium, and thorium.
b. Special nuclear material, byproduct material, or source material as defined by sections 11 a, 11e, and 11z, respectively, of the Atomic Energy Act, or any other material used in the production, testing, utilization, or assembly of nuclear weapons or components of nuclear weapons that the Secretary of Energy determines to be nuclear material under Title 10, Code of Federal Regulations, Part 1017.10(a).

When the news fails to make the distinction, the public is misinformed, which can lead to an overreaction or underreaction. A nuclear bomb involves splitting of the atom; in contrast, a dirty bomb may involve an explosion of conventional explosives and does not involve fission of the nuclear materials. Depleted uranium used in weapons is another common misunderstanding. Depleted uranium is a nuclear material. But it cannot be used for fission. And the most commonly used fissile materials—plutonium and uranium—are dangerous primarily because they can be used to make nuclear weapons. Notably, for example, 1 gram of plutonium has less than one tenth the radioactivity of 1 gram of radium. It is imperative that journalists accurately, repeatedly, and consistently report these distinctions, because fissionable material is orders of magnitude more hazardous than nonfissionable material.

Pitfalls

The media have paid too little attention to efforts to control nuclear materials. The NPT has 13 articles that establish the three pillars of nonproliferation: disarmament by nations with nuclear weapons, no nuclear weapons for nonnuclear nations, and the right to peaceful use of nuclear technology. Every 5 years, there is a nuclear nonproliferation treaty conference that discusses these principles and new ideas. In fact, there are four conferences. The first two consider procedures, the third considers format, and the last reviews substantive issues. Each of these is an opportunity for meaningful media coverage. Conferences held in 2004 and 2005 were contentious, involving debates about disarmament, lack of access by non-nuclear states to nuclear technology, and lack of assurances from nuclear states that they will not use nuclear weapons against non-nuclear states. China, for example, has stated that it will not initiate a nuclear exchange.

The United States is the focus of much of the concern about first use of nuclear weapons. These conferences and debates would appear to be extraordinarily newsworthy. However, there has been relatively little coverage in the U.S. media. A key neglected story is the diplomatic efforts and strategies required to provide sufficient security to nations that have been asked not to develop or acquire nuclear weapons. World politics has changed since the collapse of the Soviet Union, and the proliferation issue is directly tied to security as much as to economic power and growth.

There is concern today about a preemptive strike on North Korea, and about Iran using nuclear bunker buster missiles. There is very little coverage of these issues, despite the reality that these are profound issues that frighten the public and could affect the lives of tens of millions of people alive today and in future generations. In February 2007, 22 leading U.S. physicists, including many Nobel Prize winners, wrote to the U.S. Congress requesting that they restrict President George W. Bush's authority to use nuclear weapons against non-nuclear nations. Again, there has been relatively little media coverage.

The United States, the United Kingdom, China, France, and Russia belong to the nuclear weapons states. Argentina, Brazil, India, Pakistan, and South Africa began on a path to develop nuclear weapons. Three of these (Argentina, Brazil, and South Africa) discontinued their programs; while India and Pakistan developed weapons. North Korea claims to have weapons and most believe Israel has weapons. Given the implications of a nuclear exchange, the policies of each of these countries are newsworthy and deserve considerable media analysis and attention. Nevertheless, the key newsworthy issue is what the United States should be doing about nuclear weapons and disarmament.

Overall, we find that the *Guardian,* the *Nation,* Canadian newspapers, and many others do a much better job of covering proliferation than many major United States media outlets. The major shortfall is a lack of coverage of anti-proliferation.

References and Other Resources

Arms Control Association. Back issues of *Arms Control Today.* Available at *www.armscontrol.org/act.*

Braun, C., & Chyba, C. (2004). Proliferation rings: New challenges to the nuclear nonproliferation regime. *International Security,* 29(2), 5.

Bulletin of the Atomic Scientists home page. *www.thebulletin.org.*

Federation of American Scientists home page. *www.fas.org.*

International Atomic Energy Agency home page. *www.iaea.org.*

International Campaign to Abolish Nuclear Weapons home page. *www.icanw.org.*

International Physicians for Prevention of Nuclear War home page. *www.ippnw.org.*

National Academy of Sciences. (2005). *Effects of nuclear earth-penetrator and other weapons. www.tinyurl.com/388xco.*

National Academy of Sciences. Nuclear and Radiation Studies Board home page. *dels.nas.edu/nrsb.*

Nelson, R. (2001, January/February). Low-yield earth-penetrating nuclear weapons. *Federation of American Scientists Public Interest Report. www.fas.org/faspir/2001/v54n1.*

Nuclear Weapon Archives home page. *www.nuclearweaponarchive.org.*

Sokolski, H. (2001). *Best of intentions: American's campaign against strategic weapons proliferation.* Westport, CT: Praeger.

Union of Concerned Scientists home page. *www.ucsusa.org.*

United Nations Scientific Committee on Effects of Atomic Radiation (UNSCEAR) home page. *www.unscear.org.*

University of Southern California. Center for Risk and Economic Analysis of Terrorism Events (CREATE) home page. *www.usc.edu/dept/create.*

U.S. Nuclear Regulatory Commission home page. *www.nrc.gov.*

(See also state government environmental or public health agencies, radiation control sections.)

Protecting Nuclear Power Plants against Terrorism

Written by Michael R. Greenberg, based in part on an interview with Holly Harrington, with comments by Richard L. Garwin

Background

The terrorist attacks on September 11, 2001, were traumatic for the American population, business, and government. Many Americans canceled trips, especially airplane trips, and would not go to locations they perceived as likely targets; some began storing staples and other items in patterns reminiscent of the 1950s and 1960s cold war nuclear scares. After 9/11, some businesses moved parts of their headquarters and record-keeping activities away from cities and other places they perceived as likely targets.

Governments at all levels were challenged to promote internal security, none more than the U.S. Nuclear Regulatory Commission (NRC). The NRC, the watchdog agency for the U.S. nuclear power industry, promptly initiated a series of actions to enhance security both short and long term. On September 22, 2001, for example, the media reported concerns about the ability of nuclear power plants to survive a terrorist attack. Could nuclear power plants withstand airplane crashes (e.g., Boeing 757s and 767s) and survive sabotage from within or without (Schoch, 2001)? The spring 2002 issue of the *Journal of Nuclear Materials Management* focused on threats of theft and sabotage of nuclear materials and facilities. The authors (Bunn and Bunn, 2002) said that "by far" their greatest concern was an attack against a nuclear power plant or spent fuel pool at a nuclear power plant. Verbal and written statements asserted that the reactors were unsafe, that the utilities were more concerned about profits than safety, and that the NRC was ill-equipped to protect the reactors and was "in bed" with the nuclear industry. Nuclear power plants, in short, were painted as the "soft underbelly" of U.S. national security. However, by the time the spring 2002 issue was released by the *Journal of Nuclear Materials Management,* the NRC had issued dozens of security advisories and numerous orders to enhance the level of protection against both theft and sabotage. Still, in the *U.S. News and World Report,* the columnist Mortimer Zuckerman (2004) described "a nuclear nightmare" and how it must be prevented.

Before reviewing this brief, reporters should note that it offers less detail than many of its counterparts because of security concerns. Direct contact with the NRC is recommended as the best way to determine what information can be disclosed.

Identifying the Issues

In 2007, in the interview for this brief, Holly Harrington, a former journalist and currently a spokesperson for the NRC, was asked to describe NRC's efforts to address the concerns regarding terrorist attacks on nuclear power plants. (But she was not asked to assure us that there is no possibility of a successful attack.)

Harrington began by noting that NRC has always focused on safety and security, that, in fact, it was separated from the Atomic Energy Commission so there would be an agency that concentrates on regulation and has no role in promoting the business of nuclear power. Security efforts have substantially increased since the events of 9/11, she added, observing that the Nuclear Energy Institute (NEI), a nuclear energy business group, estimates that by the end of 2004, the utilities had spent in excess of $1 billion to enhance security at existing nuclear power plants. Harrington did also note that after 9/11, the NRC has been forced to withhold security-related information from the public to ensure that terrorists do not use this information against NRC licensees. In some instances this policy has inflamed the public concerns because security-related information is not available.

It should be noted that the NRC has determined that the probability of an attack cannot be reliably determined. The NRC reviews intelligence information to assess adversary capabilities and assesses the vulnerabilities and consequences associated with credible threats. Drawing on the results of those assessments, it makes decisions about the appropriate level of protection it will require of its licensees and the appropriate degree of coordination required of off-site responders.

Harrington divided her remarks between "inside the fence" and "outside the fence" activities. The NRC requires physical and human controls. Starting with physical controls, structures are hardened, that is, the reactor containment itself is a robust structure with thick concrete and steel walls. Although all nuclear power plants had barriers, checkpoints and other physical controls, and screening devices and other engineered systems before 9/11, there have been reviews and upgrades since that time in an effort to increase physical security.

One major concern is about spent fuel, which is currently being stored at all 65 active sites, at 8 sites where the reactors have been shut down, and at 1 other site. Congress, the NRC, and the Department of Homeland Security requested that the National Academy of Sciences study the vulnerability of spent fuel storage at U.S. nuclear power plants. The committee, in April 2005, examined a large number of attack scenarios. It concluded that successful attacks on spent fuel pools are possible but would be difficult to achieve. A zirconium cladding fire (fuel pellets are stored in the cladding) could release large amounts of radioactive materials. The committee recommended that spent fuel be relocated within the spent fuel pools to even out the temperature in the pools and that alternative means of cooling spent fuel pools be provided. The committee observed that it would be quite difficult for terrorists to steal spent fuel from the pools for the purpose of using the material as "dirty bombs." (See "Dirty Bombs.") As a result of the study and other research performed by the NRC, the NRC directed all licensees to spread out spent fuel assemblies around spent fuel pools. It is important to note, Harrington pointed out, that the likelihood of a spent fuel fire drops significantly, and eventually reaches almost zero, in the months following the end of cycle for that reactor assembly.

With regard to the reactor core, a nuclear reactor is not a nuclear bomb. Bombs tightly pack highly enriched fissionable material in a small area. Reactors use low-enriched uranium and a distributed structure, which will not allow a nuclear weapon explosion. The reactor is surrounded by multiple containments. The first is the metal cladding around the fuel pellets. Then a steel vessel about 5 inches thick surrounds the reactor. The building housing the reactor is typically a steel shell and 4–6 feet of reinforced concrete. The reactors are water cooled. And coolant can reach the reactor from multiple paths. Since the TMI event, the NRC has required redundancy for any safety system. In short, engineered systems to protect against terrorism are not infallible, but have certainly substantially improved since TMI. The TMI and Chernobyl problems were as much or even more human error than they were physical failures. NRC and the utilities have upgraded personnel selection requirements and training. Nevertheless, the NRC and the utilities and the press have much to learn from events at nuclear plants that do not result in injury, such as that on February 26, 2008, when two reactors of the four at Turkey Point, south of Miami, shut down automatically in response to an equipment malfunction in a substation near Miami (NRC, 2008). With regard to security, the most visible upgrade is probably the significant physical security structures that have been added to every site. In addition, each facility is required to prove that it can thwart a set of attacks. The NRC previously relied on security staffs who were not working at the specific time to play the role of insurgents. These ad hoc terrorists have been replaced by professional adversaries who are used during exercises evaluated every 3 years by the NRC staff and conducted annually by the licensee. During the NRC inspection, if the intruders are successful, the NRC does not leave the site until the vulnerability has been addressed. The staged terrorist events can include truck bombs, assault with automatic weapons, active insiders, explosives, and more. On January 29, 2007, the NRC amended a new design basis threat (DBT), which provides a general description of the types of attacks and capabilities of the attackers that the licensees must be able to protect against. Although this new DBT increased the requirements that licensees must meet, the specifics remain classified to ensure that terrorists are not able to use this information against the sites.

While the threat environment is certainly different in a post 9/11 world, the NRC has had a threat assessment branch for the past 30 years. This branch works closely with the federal intelligence and law enforcement communities, such as the FBI, the CIA, the U.S. Customs Service, the Defense Intelligence Agency, the Departments of Defense, Energy, and Homeland Security, and others, to examine evolving terrorist capabilities and credible leads about terrorist threats to plants and nuclear materials. It monitors these intelligence sources to ensure that NRC licensees are promptly made aware of any credible threat and that NRC security policies are consistent with the most current threat information.

During 2007, NRC added an explicit element for a cyber attack, an attack on computer systems. Furthermore, the NRC commissioners agreed that it should not be the responsibility of the utilities to provide active defenses against aircraft attacks and missiles. Such defenses are the responsibility of the military. The vulnerability to aircraft crashes has been a major issue, especially since 9/11. In a series of articles for the *New York Times,* Matthew Wald (2006a, 2006b, 2007) focused on NRC's recent deliberations about protecting nuclear power plants against terrorists who would try to fly an airplane into a reactor or spent fuel pool (see NEI, 2002, for a rebuttal to the claim; see also Associated Press, 2007; Chapin et al., 2002). NRC commissioner Gregory Jaczko's position is that the NRC should require new plants to be designed to withstand commercial aircraft crashes, major fires, and explosions. The NRC considers not only its own regulatory issues about aircraft but also the considerable actions of other federal agencies to secure airplanes, such as locked cockpit doors, additional screening at airports, federal marshals on the aircraft, and more. Also, the NRC has implemented additional measures at all nuclear power plants to reduce the likelihood of aircraft impact resulting in adverse public consequences.

Harrington explained that the Federal Emergency Management Agency (FEMA) is responsible for emergency preparedness outside the plant fence. FEMA works closely with site, state, and local officials to prepare a plan in the event of a real problem. The plan is regularly evaluated by staff and tested in exercises. Every site has an "emergency planning zone" (EPZ) of 10 miles surrounding the nuclear plant. NRC assumes that an airborne release might cover the entire EPZ, or more likely a wedge of it. Depending on the weather and the land use of the area where the plume is heading, evacuation or shelter-in-place could be ordered. Gregory Jaczko (2007), commissioner of the NRC, notes that NRC has sponsored research (at the Sandia National Laboratory) to try to determine under what conditions people should be sheltered in place rather than evacuated. He emphasizes that clear communications with surrounding state and local governments and with first responders is essential to avoid panic. Jaczko adds that the NRC is adding exercises based on the assumption of simulated releases.

The NRC and utilities provide recommendations regarding shelter in place or evacuation, but the state governors or, for some licensees, local officials, with the advice of public health officials, make the decision on protective actions. The governor or local official may consider the issue of potassium iodide (KI) as a supplement to protective actions, which blocks absorption of radioactive iodine, thereby reducing the probability of thyroid cancer. Not all states issue KI to the public. State officials also must advise farmers about sheltering their livestock.

What Reporters Need to Know

Some information about nuclear power plant security is classified. Specific security orders and requirements, for example, are not available to the public. Yet, reporters can request information from the NRC and the local utility, and in some cases nonclassified information will be provided and a journalist can be taken on a site tour. The same is true of FEMA, and its outside-the-fence emergency planning and response role. Some state and local security-related planning and response information may be classified, but public information and actions on what to do in the event of a nuclear power plant emergency is provided in advance to people near all nuclear plants and updated routinely. Harrington emphasized that access to information is site specific.

She noted that journalists can consult the Web sites of the Union of Concerned Scientists (*www.ucsusa.org*) and the Project on Government Oversight (POGO) (*www.pogo.org*), two organizations that have closely monitored NRC and critiqued its performance. The Union of Concerned Scientists (2007) identifies itself as "the leading watchdog on nuclear safety" and says that its "actions have resulted in safety regulations being upgraded, plants being shut down, and important modifications made to plant emergency systems and procedures." POGO (2002, 2007) focuses on worker-related issues that it believes could undermine plant safety, such as worker training, mock attacks, whistleblower protection, and worker stress.

Misunderstandings

As a former journalist, Harrington is sensitive to both misunderstandings and pitfalls. She mentioned common misunderstandings about the different kinds of radiation, the distinction between nuclear weapons and nuclear power, differences between the Three Mile Island and Chernobyl events, and the need to place radiation dose in context of background levels. These are described in the following briefs: "Radionuclides," "Three Mile Island and Chernobyl," and "Managing the Nuclear Weapons Legacy."

Harrington concentrated on one misunderstanding, which involves confusion about the NRC's role compared to the Department of Energy's. NRC was created by Congress to separate the marketing and safety issues inherent in nuclear power. NRC's responsibility is safety and security of commercial nuclear power. It is responsible for licensing and assuring the public that licensees are operating within the requirements of their license. The Department of Energy and the Department of Defense are responsible for nuclear weapons and the waste from nuclear weapons. The NRC does not manage nuclear weapons or waste from weapons programs.

Pitfalls

Harrington noted that the NRC is an independent federal agency that strives to be transparent to the public and all stakeholders. It encourages inquiries, and when there are problems it holds hearings and publishes the results. Sometimes this transparency leads to magnification of perceptions about risk and concern about NRC's capacity to manage licensees. Harrington invites members of the press to directly contact NRC staff at any of its four regional offices or headquarters in Rockville, Maryland, to request information or an interview rather than simply to repeat what someone else has said about an event. The agency's Web site (*www.nrc.gov*) is also a good source of information.

A final reminder: the contents of this brief were heavily influenced by the interview with Harrington, who is employed by the NRC. For journalists seeking to pursue this topic further, Richard Garwin, the renowned scientist who added comments to this brief, recommends *Making the Nation Safer: The Role of Science and Technology in Countering Terrorism* (NRC, 2002). See, in particular, chapter 2, "Nuclear and Radiological Threats."

References and Other Resources

Associated Press. (2007, January 30). Panel rejects responsibility for attacks. *New York Times. www.nytimes.com/2007/01/30/washington/30brfs-reactor.html.*

Bunn, M., & Bunn, G. (2002, Spring). Strengthening nuclear security against post–September 11 threats of theft and sabotage. *Journal of Nuclear Materials Management,* 1–12.

Chapin, D., Levenson, M., Pate, Z., & Rockwell, T. (2002). Nuclear power plants and their fuel as terrorist targets. *Science,* 297, 1997–99.

Jaczko, G. (2007, July 10). *Perspective on preparedness for radiological terrorism. www.nrc.gov/reading-rm/doc-collections/commission/speeches/2007/S-07-036.html.*

Nuclear Energy Institute (NEI). (2002, December). *Deterring terrorism: Aircraft crash impact analyses demonstrate nuclear power plant's structural strength.* Report prepared by the Electric Power Research Institute. *www.nei.org/resourcesandstats/documentlibrary/safetyandsecurity/reports/epriplantstructuralstudy.*

Project on Government Oversight (POGO) home page. *www.pogo.org.*

Project on Government Oversight (POGO). (2002). *Nuclear power plant security: Voices from inside the fences, September 12, 2002. www.pogo.org/p/environment/eo-020901-nukepower.html.*

Project on Government Oversight (POGO). (2007). *POGO urges NRC to approve new protections for power plant security officers. www.pogo.org/p/homeland/ha-070301-nrc.html.*

Schoch, D. (2001, September 22). Federal regulators reviewing security at nuclear power plants. *Los Angeles Times. www.commondreams.org/headlines01/0922-02.htm.*

Union of Concerned Scientists home page. *www.ucsusa.org.*

Union of Concerned Scientists. (2007). Nuclear safety. *www.ucsusa.org/nuclear_power/nuclear_power_risk/safety.*

U.S. Nuclear Regulatory Commission (NRC) home page. *www.nrc.gov.*

U.S. Nuclear Regulatory Commission (NRC). Committee on Science and Technology for Countering Terrorism. (2002). *Making the nation safer: The role of science and technology in countering terrorism.* Washington, DC: National Academies Press. Available at *www.nap.edu/catalog.php?record_id=10415.*

U.S. Nuclear Regulatory Commission. (2007). *Threat assessment. www.nrc.gov/security/threat.html.*

U.S. Nuclear Regulatory Commission (NRC). (2008, February 26). *Statement on Turkey Point Nuclear Power Plant. www.nrc.gov/reading-rm/doc-collections/news/2008/08-037.html.*

Wald, M. (2006a, November 9). Agency considers A-plants vulnerability. *New York Times.*

Wald, M. (2006b, December 25). Nuclear firms request rules to combat aerial attacks. *New York Times.*

Wald, M. (2007, April 25). U.S. takes step to address airliner attacks on reactors. *New York Times.*

Zuckerman, M. (2004, September 19). Stopping a nuclear nightmare. *U.S. News and World Report.*

International Agencies and Policy

Written by Karen W. Lowrie, with review by Frank L. Parker and comments by Richard L. Garwin

The International Atomic Energy Agency

On October 1, 1957, the United Nations created the International Atomic Energy Agency (IAEA) to be headquartered in Vienna, Austria. Its creation was a response to widespread fears about nuclear technology and its association with destructive uses. The IAEA's mission is to promote the peaceful use of nuclear energy and prevent the spread of nuclear weapons in the world. The agency was set up as the world's center of cooperation on nuclear issues, working with 81 member states and other organizational partners in three main areas: Safety and Security (development of safety standards for nuclear facilities); Science and Technology (peaceful applications of nuclear technology); and Safeguards and Verification (inspections of nuclear facilities to ensure peaceful use). The IAEA is sometimes referred to as the "UN's nuclear watchdog," although it is not under the direct control of the UN. The agency's safeguards are activities to verify that a country is abiding by international standards related to not using nuclear programs for weapons purposes. Each agreeing state must declare its nuclear materials, facilities, and activities, and IAEA verification ensures that these declarations are correct and complete.

The IAEA, with a staff of 2,200 from more than 90 countries, is led by a director general with direction from a board of governors and the general conference of all member states. Its relationship with the UN is regulated by a special agreement and by statute, which requires an annual report to the UN General Assembly on matters related to noncompliance and international peace and security. The IAEA also supports research centers and scientific laboratories in Monaco; Trieste, Italy; and Vienna and Seibersdorf, Austria.

Despite a difficult political and technical climate in the early years of the IAEA, the agency achieved progress in many areas. It created a marine environment laboratory to study the effects of radioactivity in the oceans, and it worked with the United States and the USSR to seek common ground in nuclear arms control in the aftermath of the 1962 Cuban Missile Crisis. However, in the area of nuclear safeguards, the statute and authorities of the IAEA soon became inadequate to deter fears about proliferation. Therefore, the Nuclear Nonproliferation Treaty (NPT) was approved in 1968. It was an international, legally binding commitment to stop the spread of nuclear weapons beyond the five declared weapons states at that time (United States, USSR—now the Russian Federation, United Kingdom, France, and China) and work toward the eventual elimination of nuclear weapons. The NPT was initially accepted by most industrial countries and a majority of developing countries, and now all but a handful of countries are party to the agreement, some of whom have or are perceived to have nuclear weapons.

In the 1970s and 1980s, the IAEA's functions with regard to commercial nuclear power became very important. The Three Mile Island accident and the Chernobyl disaster persuaded governments to strengthen the IAEA's role in nuclear reactor safety.

The climate for possible nuclear conflict chilled substantially in the early 1990s with the end of the cold war. At the same time, the revelation of a secret nuclear weapons development program in Iraq in the early 1990s and verification challenges in North Korea underscored the importance of a strong safeguards system to detect such undeclared nuclear activities. In 1995, the NPT was made permanent, and the UN put forth a comprehensive test ban treaty in 1996. The IAEA can now legally enforce additional protocols, which are expansions of the state's original safeguards agreement to cover all aspects of nuclear fuel–related activities, from research to mining to disposition. The IAEA took on duties such as verification of the peaceful use and storage of nuclear materials and surplus military stocks, and verifying safety of some former nuclear test sites and wastes from sunken warships. In the 21st century, the IAEA has added nuclear terrorism countermeasures to its action plan. However, to say that the IAEA can now "legally enforce additional protocols" may mislead some readers. Indeed the IAEA can deploy inspectors to a country that has signed the additional protocols, and they have substantial rights of travel, inspection, and reporting. But if the IAEA finds that a country is not living up to its additional protocol, its sole recourse is to report this lapse to the U.N. Security Council, which, in turn, may or may not take action.

Today over 30 countries have nuclear reactors, and over 70 countries have major facilities with some type of nuclear material that is "safeguarded" under IAEA agreements with governments. The ultimate success of the impartial inspectorate system depends on the continued support of the international community, the openness and honesty of the member states to submit to transparency in nuclear activities, and the extent to which inspectors have access rights to perform independent verification.

The Nuclear Energy Agency (NEA)

The NEA, a specialized agency within the Organization for Economic Co-operation and Development, has the mission "to assist its Member countries in maintaining and further developing, through international co-operation, the scientific, technological and legal bases required for the safe, environmentally friendly and economical use of nuclear energy for peaceful purposes." The NEA collaborates with the IAEA, complementing its work in the field of radioactive waste management to promote the adoption of safe and efficient policies and practices. The agency's studies focus on the cost, feasibility, and safety of radioactive waste disposal alternatives and decommissioning methods. The Radioactive Waste Management Committee reviews progress on waste disposal strategies and policies in member countries from around the world. Results of NEA studies and cooperative programs are disseminated to the international nuclear community.

Other International Policy Concerns and Initiatives

It is important to note that international situations are fluid, and while we summarize the issues as of the time of publication of this handbook, the reporter will need to examine reliable sources to find updates to these evolving issues.

The United States and India

India is not part of the nonproliferation treaty, and hence it has largely not been part of international trade in nuclear materials. Because of its need to add electricity generation capacity, India has been moving toward using native thorium deposits to build a set of nuclear power plants. In October of 2008, President George W. Bush signed the United States–India Nuclear Cooperation Approval and Non-proliferation Enhancement Act to strengthen the partnership between the two countries to meet energy and security challenges and thus help India to meet its growing demands for electricity with nuclear energy, easing its demands for fossil fuels and reducing its emissions of greenhouse gases. In August 2008, the IAEA agreed to bring India's civilian nuclear program under the safeguards of the IAEA after 30 years outside the system, in exchange for access to U.S. technologies.

Iran

Inspectors from the IAEA reported in 2003 that Iran had for 18 years repeatedly failed to meet its safeguards obligations, including failing to declare its uranium enrichment program. Later that year, Iran pledged to implement an additional protocol and to suspend its plutonium reprocessing and uranium enrichment-related activities. However, Iran ceased implementation in 2005, leading the IAEA to rule that Iran had failed to meet its safeguards obligations. After an IAEA report to the UN on the matter, the Security Council passed a resolution for sanctions against Iran until it suspends enrichment activities. However, the sanctions are currently on "time out" to encourage negotiations. In early 2007, Mohamed El Baradei, director general of the IAEA reported that there was a continuing stalemate in verifying Iran's nuclear program. In late 2007, the IAEA verified the nondiversion of declared nuclear material but not the absence of undeclared activities.

North Korea (Democratic People's Republic of Korea)

The uncertain status of North Korea's nuclear program has caused growing suspicions and fears about its nuclear weapon capabilities in recent years. North Korea entered into the NPT in 1985 and into its Safeguards Agreement in 1992. However, beginning in the early 1990s, IAEA inspectors found inconsistencies between North Korea's declarations of certain plutonium and other waste products and the agency's own analysis, suggesting the presence of undeclared plutonium. Throughout the 1990s, various charges of noncompliance were raised and continuing struggles for access and verification between the IAEA and North Korea existed. An agreed framework was negotiated in 1994, which allows Korea to pursue a new light water reactor (under the Korean Peninsula Energy Development Organization) in exchange for freezing and dismantling its older graphite-moderated reactors. But by 2002, charges of a program to enrich uranium for nuclear weapons in North Korea implied violations of the NPT, safeguards agreements, and the agreed framework.

The administration of George W. Bush initially took a principled approach that did not permit bilateral negotiations with North Korea—an approach assiduously forwarded by John Bolton, first as a State Department official and then as the U.S. ambassador to the United Nations. But eventually in 2007 the Bush administration became amenable to direct discussions. The Bush administration announced that North Korea had admitted operating a program to enrich uranium, unreported to the IAEA, but North Korea has made no public statement to this effect.

The actions of North Korea have caused continued concern in recent years, and the country withdrew from the NPT effective in early 2003. The UN expressed concern over the situation, and the IAEA maintained that a continued dialogue was important to achieving a peaceful resolution. The UN has urged North Korea to continue its participation in the six-party talks (with representatives from North Korea, China, Japan, Korea, Russia, and the

United States), particularly in response to ballistic missile launches and a reported nuclear test in 2006. The UN Security Council adopted a resolution in October 2006 imposing sanctions against North Korea until it ceases development of weapons and returns to the NPT and submits to IAEA safeguards. In July 2007, an IAEA inspection team visited the country to verify the shutdown of its Yongbyon nuclear facility, pursuant to an invitation from North Korea to visit for talks. Surveillance and monitoring equipment placed at the facility will help to ensure compliance and is a first step toward better international cooperation on the part of North Korea.

Expert Group Report on World's Nuclear Fuel Cycle

An international expert group with representatives from 26 countries examined issues related to the world's civil nuclear fuel cycle and in early 2005 released a report called "Multilateral Approaches to the Nuclear Fuel Cycle" (IAEA, 2005). The report cites six approaches to strengthen controls over proliferation sensitive nuclear materials and technologies.

The approaches, summarized, are as follows:

1. Reinforcing existing commercial market mechanisms through long-term contracts and transparent suppliers' arrangements with government backing. Examples would be fuel leasing and fuel take-back offers, commercial offers to store and dispose of spent fuel, and commercial fuel banks
2. Developing and implementing international supply guarantees with IAEA participation
3. Promoting voluntary conversion of existing facilities to multilateral nuclear approaches
4. Creating voluntary agreements for new facilities based on joint ownership, drawing rights, or co-management for uranium enrichment; fuel reprocessing; disposal and storage of spent fuel (and combinations thereof)
5. Developing a nuclear fuel cycle with stronger multilateral arrangements for expanding nuclear energy—by region or by continent—and for broader cooperation, involving the IAEA and the international community.
6. Replacing HEW (heavy water) with LEW (light water) for research reactors.

Misunderstandings

The IAEA performs regular inspections under its safeguards program of all states with declared materials or facilities. These inspections, however, are different and distinguished from special inspection responsibilities that may be assigned to deal with special circumstances under specific UN Security Council resolutions. An example is Resolution 687 (1991), which charges the IAEA to develop a special team to uncover and dismantle Iraq's nuclear program. The implementation of the activities of such specialized teams is reported directly to the UN Security Council by the IAEA's director general.

Pitfalls

The media need to be thorough and balanced in their coverage of the international nuclear cooperation proposals. For example, what are the pros and the cons of the Global Nuclear Energy Partnership (GNEP)? (See "Closing the Civilian Nuclear Fuel Cycle.") Should the United States take measures to reprocess its existing spent fuel? Will a revival of nuclear power remove the incentive for conservation worldwide?

Along with discussion of nonproliferation, decommissioning nuclear weapons, and civilian spent fuel briefs, this brief provides context for current and near-term controversies. But it comes with a caveat: Because of the ability to transfer technology and knowledge, it is not possible to say with certainty where the biggest controversies will be found in a decade.

References and Other Resources

International Atomic Energy Agency home page. *www.iaea.org.*

International Atomic Energy Agency (IAEA). (2005). *Multilateral approaches to the nuclear fuel cycle. www.iaea.org/NewsCenter/News/2005/fuelcycle.html.*

Nuclear Energy Agency home page. *www.nea.fr.*

Section 5
Risk Perception and Risk Communication

Global Warming and Fuel Sources

Written by Michael M. Greenberg, based in part on an interview with Paul Meier, with comments by Sandra Quinn

Background

This discussion of global warming is in a section about perception and communication because global warming has become a prominent political issue that many politicians use in their public statements to tie together nuclear power and climate change. Carbon dioxide (CO_2) is a heat-trapping gas that acts like a blanket, thereby keeping the earth warm enough to support life. The earth has been trapping solar energy and building carbon-bond fossil fuels from the decay of organic material for hundreds of millions of years. However, humans have been rapidly consuming fossil fuels, much of it during the past century. The oceans, soils, and plants absorb CO_2 out of the atmosphere, but CO_2 buildup has been faster than these natural processes can accommodate. As CO_2 increases in the atmosphere, the planet is getting warmer. Warming of the planet by additional CO_2, methane, ozone, nitrous oxide, and water vapor (not clouds) is called "global warming." This brief summarizes the issue and comments on the likelihood that nuclear power can substantially contribute to curbing greenhouse gas emissions.

The consumption of fossil fuels for energy production is the consensus primary cause. The Energy Information Administration reported that in 2005, over 86% of primary energy was produced by petroleum (36.8%), coal (26.6%), and natural gas (22.9%). The remainder was produced by hydroelectric power (6.3%) and nuclear electric power (6.0%); geothermal, solar, wind, waste, and wood were responsible for 0.9%. Global energy demand has grown rapidly and will probably continue to grow because energy use has been fundamental to growing populations and growing economies.

What Are the Risks?

Increased storm intensity, changing precipitation patterns, and increased heat are some of the symptoms of global warming. Urban areas along coasts and major rivers face increased flooding. Infrastructure is undermined and damaged by storms and erosion, and agricultural areas could have too much rain or not enough rain. Heat stress in urban areas is deadly to vulnerable populations, as demonstrated in Chicago in 1995 and Paris in 2003. Certain regions of the developing world are particularly susceptible to famine, water shortages, and disease outbreaks.

The Intergovernmental Panel on Climate Change (IPCC) is considered the authoritative source on the science and impacts of climate change. Established in 1988 by the World Meteorological Organization and the Environmental Programme of the United Nations, IPCC prepares assessments based on the weight of the scientific evidence. IPCC includes representatives from many nations, which most observers consider a major strength, but some think slows down and tempers their work. IPCC published reports in 1990 (with a 1992 supplement), 1995, 2001, and 2007. The 2007 report, like its predecessors, is voluminous—notably, the observations appear to be less equivocal than those in the earlier reports. For example, the majority of the global increase in temperature during the past half century is directly attributed to anthropogenic greenhouse gas accumulations. World temperature, the 2007 report notes, is likely to rise between 1.1° and 6.4°C (2.0° and 11.5°F) during this century. Sea levels are likely to rise by 18 to 59 cm (7.08 to 23.22 inches). The planet will experience more frequent warm spells, heat waves, and heavy rainfalls. Tropical storms are likely to increase in number and intensity, along with high tides. Also, droughts will be exacerbated, along with the continued melting of glaciers and ice caps. IPCC observes that the current greenhouse gas blanket will produce impacts for the duration of this century, even if greenhouse gas levels were to be quickly stabilized, which seems highly unlikely. In a few decades, IPCC expects tens of millions of people will be hungry because of changing climatic conditions (added to the many millions that are already suffering from hunger). Millions will face an increasing frequency of floods, tornadoes, hurricanes, and many more will face drought conditions. In the United States, conditions will not be as serious, but compared with today's environment could be much more difficult for people who live in low-lying areas, people who farm, and people whose livelihoods depend on predictable weather patterns. Many who already suffer seasonal allergies will face longer pollen periods, thereby increasing their risk of asthma, nasal and sinus infections, and related syndromes.

For those who prefer to see the big picture in a table or graph, IPCC prepared a table that classifies 19 impacts of climate change on tourism, water resources, vector-borne diseases, wildlife, and other attributes for Africa, Asia, Latin America, Australia and New Zealand, Europe, North America, the polar regions, and island nations. Information in the 152 boxes (8x19) is scored on a 5-point scale from strongly positive to strongly negative impacts. Twenty-one of the 152 are classified as "no information." Seventeen (11%) are positive impacts primarily due to warming of largely polar areas that would be more able to support agriculture, forestry, and other human activities, as well as provide potable water and transportation to areas often inaccessible because of extreme weather conditions.

In contrast, 73% (111 of the 152 impact cells) are classified as "negative" or "strongly negative." Africa and island nations, in particular, are expected to suffer from flooding, destruction of marine ecosystems, and loss of agriculture and other local attributes.

What Reporters Need to Know

A reporter can take a story about global warming and fuel sources in a variety of directions. Each direction demands somewhat different information. The health, environmental, economic, social, and political risks of global warming, as noted earlier, are becoming clearer but require continuing media coverage of global, national, state, and local levels. IPCC should be a central repository for information, but some state governments are starting to develop their own information about impacts and potential impacts.

According to the Energy Information Administration, in the United States nearly 85% of anthropogenic greenhouse gas emissions are from the burning of fossil fuels, with an additional 10% resulting from methane leakage from coal mining and gas production and transport. Other notable sources include hydrofluorocarbons, nitrous oxide, and other man-made gases.

Management of the energy alternatives is central to managing the global warming problem, and there are only three fundamental energy-related alternatives: reduce energy use, capture carbon emissions, and change the fuel supply to non-carbon-based sources, such as nuclear power or renewables.

The presentation that follows is an overview of some of the ideas being considered. Journalists will need to consult the sources listed at the end in "References and Other Resources" for more detailed presentations; for example, America's Energy Future: Technology Opportunities, Risks, and Tradeoffs (n.d.).

Reducing Energy Use

Utilities can reduce energy use by developing a "smart grid." A smart grid optimizes the location of meters and switches so that the system functions more efficiently, not wasting energy. Energy use can be reduced by capturing the one half or two thirds of the energy from utility facilities that would otherwise be lost. These cogeneration facilities can be helpful, but more research and development is required to make them more effective than they currently are. The automobile industry can build more effective hybrid cars that can use a variety of fuels, including gasoline, ethanol and hydrogen fuel cells. Vehicle emissions are a product of miles traveled, vehicle efficiency, and fuel carbon intensity. Automobiles can be designed to emphasize high-efficiency internal combustion and minimize drag. Additional efficiency gains are possible in hybrid electric vehicles by converting kinetic energy (via regenerative braking) into electrical energy, storing that electrical energy and feeding it back into the power system of the vehicle.

The federal and state governments will have to provide tax benefits for research and development of progressive technologies, and doubtless tax relief or other incentives to companies and local governments that heavily invest in these technologies. State and local governments have important roles to play. They will need to review and change their building codes, zoning, and other local practices that otherwise hinder green buildings and discourage redevelopment and the use of recycled products. Developing "smart growth" plans means concentrating development in already urbanized areas and discouraging low-density sprawl. State and local governments will need to expand local infrastructure, which will be expensive and doubtless require a subsidy from state government. (Other infrastructure, such as water, sewer, roads, schools, and other facilities would also need to expand.)

Local government and business need to introduce the principles of "green building" to development and redevelopment. This means using solar panels, recirculating and reusing water, recycling wood and other building products, and in many other ways emphasizing the use of low-energy-demanding building products and activities. Government and business will need to emphasize "progressive" purchasing practices, such as phasing in hybrid vehicles. And local governments can help by building walking and bicycling paths and by extending public transit opportunities.

Capturing Carbon Emissions

Electricity-generating stations that use coal emit a sizable and rapidly increasing share of greenhouse gases in the atmosphere. IPCC (2005) asserts that carbon dioxide from these and other sources can be captured, transported (by pipeline or other means), and stored underground in geological repositories on land and in the oceans. For long-distance shipment (1,000 km or more), pipelines would be required. Shorter distance storage sites could be reached with trains, large ships, and offshore platforms with pipelines reaching the repository.

Because of the heavy reliance on coal-powered electricity generation, carbon capture and storage (CCS) has major potential to mitigate atmospheric greenhouse gas emissions. Capture, transport, and geologic injection of CO_2 have been demonstrated technically, but large-scale economic development and the permanence of geologic storage are not yet proven. The technology is more economical when used in integrated gasified coal combustion (IGCC) facilities, allowing pre-combustion carbon removal. The vast majority of existing and pending coal facilities are not IGCC facilities and instead use pulverized coal technology, which requires more energy-intensive and expensive postcombustion separation of CO_2.

Clearly, capturing and sequestering carbon is a massive technological and logistical undertaking. In addition to carbon capture and storage, natural sinks of carbon can be increased through land management. Adding nutrient to the ocean has also been proposed.

Changing Fuel Supply

Changing the fuel supply to nuclear or renewable sources is in progress, but many feel not rapidly enough. With regard to satisfying baseload needs, nuclear fuel seems like the most obvious choice for the next decade, at least. Even if fossil fuel emissions from mining, construction, and transportation related to nuclear power are counted, CO_2 emissions from nuclear power are negligible.

In conjunction with hybrid electric vehicles that would be connected to the electricity grid, nuclear power could provide clean transportation fuel. The fossil fuel economy works by shoveling, pumping, and by other means removing carbonaceous materials from the ground, refining them, and then transporting them so that they can be burned. The hydrogen economy uses electricity to create hydrogen by splitting water into pure hydrogen and oxygen by electrolysis of water. Hydrogen also can be captured from fossil fuels by reforming hydrocarbons. However, reforming hydrocarbons would emit carbon to the atmosphere, unless substantial resources were expended to capture and use it. Consequently, reformulation from fossil fuels is, at best, a short-term solution.

Electrolysis of water would produce hydrogen fuel to power automobiles. Realistically, only nuclear and solar-based technologies appear likely to provide sufficient energy to operate a hydrogen economy. Solar energy is collected from sunlight. Electricity is produced by using concentrated solar power, by producing photovoltaic solar cells (PV), and by heating enclosed (trapped) air that turns turbines. Solar technology can heat or cool water for air conditioning, heat food using solar ovens, and heat or cool buildings with passive well-located solar collectors.

The amount of solar energy that falls on the earth is orders of magnitude greater that the energy currently consumed by humans. Yet, the challenges of using solar power as a source of energy are daunting. Current solar collectors require a considerable amount of space. Sunlight varies by location, season, and time of day, which is an issue of storage and reuse of captured energy potential. Improved systems will require substantial investments in cell construction, collection and distribution systems, automobile engines, and methods of storing solar energy and of transporting hydrogen fuel.

Technically, these do not appear to this writer to be impossible barriers. Institutional barriers seem more challenging. Where will research and development funds come from to support breakthroughs? Will large companies that have been committed to fossil and nuclear fuels actively oppose these investments? Will state and local governments modify their building practices to encourage solar power or the hydrogen economy, or will they oppose it because it can change the way they are used to dealing in the economy?

A large increase in nuclear power faces many obstacles but probably is less challenging to existing organizations. Nuclear power is described elsewhere in this volume (see "Sustainability," "Closing the Civilian Nuclear Fuel Cycle," "Nuclear Power Plant Safety," and "Public Perception").

So, can nuclear power be a major contributor or only a marginal player in worldwide efforts to reduce greenhouse gases? If the nuclear fuel supply is plentiful, if countries are able to build plants faster than they have in the past, if the skilled work force expands in response to need, if waste is successfully managed, if fossil fuel costs accelerate and the costs of other options remain high, if fuel cycle security is successful, and if national governments choose to favor nuclear power over other options and cooperate with each other to build an international program, then the impact on emissions compared with continuing current practices could be measurable in a decade. Each of these assumptions is reviewed in this volume, and even though we cannot presume to know which, if any, is the most problematic, reporters can focus on these assumptions to direct questions at policy makers.

Space does not permit a full accounting of other options. Wind, tides, geothermal heat, biomass, and other renewable energy sources that now contribute a small amount of energy do make a major contribution in some locations, and those contributions should increase. Strong winds blowing off shore and through valleys can turn wind turbines. When combined with other resources and storage systems, winds can be valuable renewable sources of energy. However, anyone who has seen the massive propellers is aware of their aesthetic impact and potential ecological effects on bird and other nonhuman species. Hydro power captures energy from rapidly flowing water, tides, and waves and uses it to generate electricity. Yet, there is consistently strong opposition to drowning valleys because they often have some of the most attractive vistas and contain irreplaceable cultural artifacts.

Plants use photosynthesis to produce biomass, which can be harvested and used to produce energy. Corn, vegetable, and animal fats, sugar beets, and other liquid and solid biomass, as well as biogas from waste can be burned in boilers and engines. However, these require energy to grow the biomass, and in the case of corn we have observed that using ethanol for automobile fuel means less corn for cattle, which implies higher prices on agricultural products. The earth's crust can be tapped to extract geothermal energy. Finding appropriate sites and dealing with aesthetic and environmental impacts of geothermal energy is an issue. Cost is a constant concern across all of these non-fossil-fuel technologies. Some, such as hydroelectric and geothermal, are relatively well developed, but others, such as solar, are not.

Before leaving this topic, which has emphasized turning to renewable energy sources, we take a brief look at coal. When asked about forces that will reduce coal use, most immediately think about global warming legal provisions. In the United States, these provisions have relatively little political teeth. The Clean Air Act, however, does have provisions related to fine particulate emissions that could require U.S. utilities to re-evaluate reliance on coal. Bernard Goldstein, one of the cofounders of the Consortium for Risk Evaluation with Stakeholder Participation, estimates that almost 90 million U.S. residents (about 3 of 10) live in areas that exceed the fine particulate standard. Reporters who write for media in these areas will find some interesting stories about the pressure on local utilities to re-evaluate their fuel choices.

Finally, Meier (2002) observes that natural gas combined-cycle electricity generation emits about 62% of the greenhouse gases that are emitted by coal plants, from a life-cycle perspective. So natural gas is a relative benefit, at least compared to coal. But nuclear energy, wind, and photovoltaics produce less than 0.5% of the life-cycle greenhouse gas emissions per kWh when compared to coal.

Pitfalls

There is a general misconception that renewable sources provide a significant portion of energy. They may proliferate in locations where particularly strong winds, solar energy, geothermal energy, and tides and other opportunities to harness water are available. But worldwide, as noted earlier, a small fraction of our energy consumption is provided by renewable sources. The potential for renewable energy is present in specific locations, but there will be impacts that will be contested heavily by those who are near proposed sites, along transit routes, or near waste disposal sites. Nor are economic benefits and costs likely to be equitably distributed.

More generally, too much attention is being focused on the symptoms and not enough on the deep roots of global warming. One root is a rapidly growing world population, which grew from 1.6 million in 1900 to 6.1 billion in 2000. At a minimum, it appears that the world population will increase to 9 billion by the mid-21st century and could reach 14 billion. Combined with substantial population growth is a more rapid increase in much of the world population's desire to adopt lifestyles commonly associated with North America, Western Europe, and Japan. These dreams, fanned by worldwide communications, are leading to rapidly growing per capita energy demands. We need to revisit the works of and Commoner (1971) and Meadows et al. (1972), who emphasized the limitations of growth. Their doomsday predictions did not materialize. But their arguments stimulated considerable discussion in classrooms and the seats of government. The media can stimulate a much needed discussion.

References and Other Resources

America's Energy Future: Technology Opportunities, Risks, and Tradeoffs. An ongoing project by the American Council for an Energy Efficient Economy, and the National Academy of Sciences. (n.d.). *www8.nationalacademies.org/cp/projectview.aspx?key=48817.*

Commission on Engineering and Technical Systems. (1990). *Fuels to drive our future.* Washington, DC: National Academies Press.

Committee on Nuclear and Alternative Energy Systems. (1982). *Energy in transition, 1985–2010.* Washington, DC: National Academies Press.

Commoner, B. (1971). *Closing the circle.* New York: Knopf.

Energy Information Administration. (2005). World net geothermal, solar, wind, and wood and waste electric power generation. *International Energy Annual. www.eia.doe.gov/pub/international/iealf/table28.xls.*

Energy Information Administration. (2007). *International Energy Annual. www.eia.doe.gov/iea/overview.html.*

Intergovernmental Panel on Climate Change (IPCC). Working Group III. (2005). *Carbon dioxide capture and storage: Summary for Policy Makers.* IPCC Special Report. *www.ipcc.ch/pdf/special-reports/srccs/srccs_summaryforpolicymakers.pdf.*

Intergovernmental Panel on Climate Change (IPCC) Working Group I. (2007) *Summary for policymakers. www.ipcc.ch/pdf/assessment-report/ar4/wg1/ar4-wg1-spm.pdf.*

Meadows, D., Meadows, D., Randers, J., & Behrens, W. (1972). *Limits to growth.* New York: Club of Rome.

Meier, P. J. (2002). *Life-cycle assessment of electricity generation systems and applications for climate change policy analysis.* Fusion Technology Institute, University of Wisconsin. *fti.neep.wisc.edu/pdf/fdm1181.pdf.*

National Hydrogen Association home page. *www.hydrogenassociation.org.*

U.S. Department of Energy. Office of Science. (2007). *DOE Mission Focus: Biofuels. www.genomicsgtl.energy.gov/biofuels/index.shtml.*

U.S. Department of Energy. (2007, May 16). *New Hampshire enacts a renewable requirement with solar set-asides. www.eere.energy.gov/states/news_detail.cfm/news_id=10781.*

Public Perceptions of Risk and Nuclear Power, Nuclear Weapons, and Nuclear Waste

Written by Bernadette M. West, based in part on information provided by Michael R. Greenberg, with comments by Sandra Quinn

Background

Research shows that the public fears the risks of nuclear war, nuclear power, and nuclear waste more than other risks. Images of the mushroom cloud and the deaths of thousands of people following the bombing of Hiroshima and Nagasaki have been indelibly printed in the minds of the public. These fears have been compounded by events at Three Mile Island (TMI) and Chernobyl, which raised the specter of disaster at a nuclear power facility close to home. (See "Three Mile Island and Chernobyl.")

Perceptions of risk are affected by individual and group values, political process, political power, and public trust. They play an important role in the decisions people make. Negative public perceptions regarding nuclear materials have been linked to its military legacy, history of secrecy, issues of government and industry distrust, the apparent vested interests of some nuclear advocates, fear of its catastrophic potential, issues of personal controllability, fear of the unknown, concerns regarding equity, and potential impact on future generations. (See the discussion of risk in Part I, "Crosscutting Themes.")

The media plays an important role in shaping public perceptions of risks. How the public perceives risks in relation to nuclear concerns is important because public perceptions could affect critical decisions about whether to increase U.S. reliance on nuclear power and where and how to clean up and manage high- and low-level nuclear waste. This brief addresses the following questions: How does the public perceive nuclear materials? Do some nuclear concerns cause more fear than others? Have public perceptions changed in the past decade or more? Will concern over global climate change and the search for alternatives to fossil fuels cause the public to reassess the role of nuclear power in the future? And, finally, how can the media effectively convey risks—to address concerns regarding both hazard and outrage?

Identifying the Issues

Studies suggest that worries about the use of nuclear technology, such as for the construction of nuclear weapons, rank at the top of most people's lists when they are asked to rank risks. Use of nuclear power to generate electricity and storage of nuclear waste also are ranked high on people's lists. Slovic (1987) found that nuclear power was the highest ranked risk out of 30 different risks ranked by college students and members of the League of Women Voters. Studies in other countries have produced similar findings. Cha (2000) found that South Koreans ranked nuclear weapons/war, nuclear weapon tests, and nuclear reactor accidents 1, 2, and 4, respectively, among 70 risks. Disposal of radioactive wastes, transportation of nuclear materials, and nuclear power plants were ranked somewhat lower—11, 12, and 19, respectively—still in the top 20. Chileans ranked nuclear weapons number 1 among 54 risks, and nuclear power was ranked 4th (Bronfman and Cifuentes, 2003). In China, respondents ranked nuclear war number 1, but nuclear power ranked much lower—27 out of 28 risks ranked (Xie et al., 2003).

Public perceptions of risk can be different from how experts view risks. Slovic and Weber (2002) found that experts base their ratings of risks on technical metrics, such as annual fatalities, while lay people tend to focus on broader concepts, such as potential threat to future generations. Hazards, such as smoking, that produce hundreds of thousands of deaths and injuries annually are ranked much lower than hazards produced by nuclear power plants. While nuclear accidents may be low probability, they are viewed as high magnitude and therefore more frightening to people than high-probability/low-magnitude events, such as the risks associated with smoking and with drinking and driving.

Understanding the public's perception of risks and how they may differ from expert perception of risk makes it important to understand the components of risk. Covello and Sandman (2001) argue that perceived risk is a composite of hazard plus outrage. They assert that hazard—defined as magnitude times probability times the number of people exposed—is what technical experts focus on in assessing risks. Outrage, in contrast, refers to the nontechnical aspects of risk—those aspects of certain situations that people find upsetting or frightening. Research has shown that the public tends to become upset and to fear risks that potentially could affect many people, risks that are perceived as coerced rather than voluntary, risks viewed as controlled by others (such as flying) rather than ones where they feel they have control (such as driving), man-made risks (such as an explosion at a gas refinery) rather than natural risks (such as a tornado), risks that are exotic and not well understood (such as SARS) rather than familiar risks (such as smoking). Add to this the fact that people are more likely to remember negative events than positive ones—especially when such events are reinforced by media coverage. Public perception of risk becomes the complex integration of emotions, past experiences, and current knowledge into a mental model that a person builds, blending such factors as catastrophic potential, controllability, dread, equity, impacts on future generations, uncertainty, and other factors.

When technical experts discuss risk with communities, they often focus on issues related to technical risk and ignore issues of outrage. The public, however, tends to focus on outrage and give less credence to measurable risk. As a result, technical people sometimes ignore certain risks with low probability that nonetheless produce fear and considerable outrage. Sunstein (2006) refers to this miscommunication as "misfear"—humans are prone to sometimes fear in the absence of danger and, at other times, overlook important risks. Misfear can present problems for both individuals and governments.

Historically, public perceptions of nuclear power, nuclear weapons, and waste sites have been negative. Public perception of nuclear issues was first shaped in large part by the image of the mushroom clouds that formed over Hiroshima and Nagasaki. Despite efforts by the Truman and Eisenhower administrations to change this image, starting in the 1950s with the construction of the first nuclear power plants, and despite efforts by the United Nations to emphasize peaceful uses of the atom, it has been impossible to erase the image of the blast and fallout from people's minds.

More than a dozen surveys conducted around the world have shown that public perception has historically been primarily negative regarding nuclear technology and became even more so following the events of TMI and Chernobyl. When asked questions about nuclear matters, many people have an image of nuclear bombs and nuclear explosions. In almost all surveys, nuclear war and nuclear weapons are viewed as the most feared hazard—more than chemical pollution, guns, airplane and automobile travel, and cigarette and alcohol abuse. Nuclear reactor accidents, disposal of radioactive wastes, transportation of nuclear materials, and nuclear power plants are also ranked high in terms of concerns.

For example, Rosa found in 2001, and again in 2004, that most Americans were opposed to having a nuclear power plant located in their jurisdiction, fearing reactor safety and waste disposal and lacking trust in both the nuclear industry and the government. Gallup surveyed Americans in an effort to measure support for nuclear power in 2004, 2006, and 2007. When averaged, the data from the three polls show that 52% of Americans were pro nuclear power and 42% opposed nuclear power.

Some trend data suggests a possible change in public perceptions regarding nuclear-related issues. In 2005 Bisconti Research, working with NOP (formerly RoperASW) and funded by the Nuclear Energy Institute, found that while many people fear nuclear materials, there is some evidence of a growing level of acceptance, especially of nuclear power. Bisconti began tracking U.S. public perceptions regarding nuclear issues in 1983. During the first decade or so, between 49% and 55% of the population said they were in favor of using nuclear fuel to produce electricity. In 2005 they reported that the percentage increased to more than 70%. In addition, Bisconti found that the difference between strong support and strong opposition had widened considerably between 1990 and 2005. Whereas about 22% were strongly in favor and strongly opposed to using nuclear fuel to generate electricity in 1990, by May 2005, 32% strongly favored using this approach and only 10% strongly opposed it. Although the Bisconti findings are produced by an industry-sponsored research group, they suggest the possibility of a change in public perception. The survey found that in 2005, 85% of respondents agreed with renewing the license of nuclear power plants that continue to meet federal safety standards, 75% agreed that electric utilities should prepare to build new plants if needed, and 58% agreed that the United States definitely should build more nuclear power plants. Although Bisconti started asking these questions only in 2003, reported levels of support for these positions have increased. The answers to these questions imply that the American public does not want to apply the "not in anybody's backyard" (NIABY) policy to nuclear power plants. The Bisconti data also show public support for clustering nuclear power plants, that is, building new facilities in close proximity to existing plants. Over two thirds (69%) said clustering was acceptable in 2005, compared with 57% in 2003. There was some regional variation, with more support from respondents in the Midwest compared with those in the Northeast. Two thirds of respondents (64%) said there was a nuclear power plant in the state where they live, suggesting the absence of a strong "not in my backyard" (NIMBY) sentiment.

In a similar vein, Ansolabehere (2007) reported in an MIT study that close to two thirds of respondents to a survey conducted in the United States said they would favor expanding nuclear power if more effective solutions could be found for waste storage problems. In this same survey, close to two thirds of those sampled said they supported the U.S. Department of Energy's reprocessing program. Half of respondents to the survey reported support for expanding nuclear energy in the United States if fuel were reprocessed.

The findings from some of these studies seem to be at odds with the findings from others—suggesting a possible shift in attitudes, though most findings demonstrate continued strong negative sentiment. The lack of robust studies, however, makes conclusions here difficult. The shift in attitudes suggested by some of these studies may be linked to the following factors: the impact of rising oil prices, the passage of time and the fact that an increasing number of Americans were not alive or were too young to remember events such as TMI and Chernobyl, and growing concerns about global warming and the search for alternative fuels that do not contribute to greenhouse gases. Recently public perceptions regarding nuclear fears have collided with growing public concern about global climate change and how to address it. Lieter (2007) differentiates between "hot" risks (nuclear) and "cooler" risks (climate change). Perceptions about hot risks usually prove to be more resilient to shifts in public opinion.

Level of trust is an important factor in shaping public perceptions of risk. Research suggests that when people lack trust in the agencies responsible for managing an activity, very often they perceive the activity as more risky. For example, people who lack trust in the U.S. Department of Energy or the U.S. Nuclear Regulatory Commission are more likely to view the risks of nuclear power as greater. When information is presented by trusted sources, however, people often give more credence to it than to information supplied by a less trusted source.

Some people—in particular, those who are less powerful and less trusting of authority in general, including those from lower socio-economic strata and those with little knowledge about nuclear matters—are more likely to distrust the federal government and the nuclear power industry, believing that the government does not do enough to protect them, and that the nuclear industry is primarily concerned with profits. They believe neither communicates well about matters involving nuclear concerns.

What Reporters Need to Know

When nuclear materials are the story, the media needs to ask about the specific materials involved, how much is involved, the level of risk, the likelihood of exposure, and diffusion of the risk, as well as the response of officials—to avoid immediately drawing the story to fear.

Perception of risk varies according to whether the situation involves nuclear weapons, waste from weapons or from medical facilities, or other sources. To conclude that all people fear anything nuclear because they fear nuclear weapons is too simplistic. Food irradiation, for example, is a very different concern than nuclear weapons and nuclear war. There are noteworthy differences between nuclear fuel rods and nuclear waste from hospitals. (See "Civilian Uses of Radiation and Radioactive Material.")

To determine the specifics involved in any nuclear situation, reporters should ask, for example: Does it involve a nuclear weapon or a nuclear power plant? If the concern involves terrorism and nuclear weapons, what is the nuclear material? Does the weapon involve fission? If it does not, the reporter needs to make clear how the effects are very different. For example, in covering a story involving a dirty bomb, the reporter needs to make it clear that dirty bombs do not involve fission and therefore the consequences are very different from those involving the detonation of a bomb involving fission. (See "Dirty Bombs.") If it is a situation that involves a nuclear power plant, is it one located in the local area or 300 miles away? Is the issue one involving storage of low-level nuclear waste? If it is a transportation concern, is it transportation of medical waste or nuclear fuel rods? There are some nuclear materials from which the exposure dose is so small it is highly unlikely they will cause any damage, and then there is nuclear material that is very hazardous. The devil is in the details, especially because of the perceptual implications.

Reporters need to investigate the differences, if any, between the perceptions of people who live near, and work at, the existing site and people who live further away. There are clearly differences based on location that are predictable, based on what would be expected from the risk perception literature. Studies suggest there is variation in public perceptions regarding risk based on level of familiarity with the issue at hand. On one hand, people who benefit directly—for example, those who work at a nuclear site or have a friend, relative, or neighbor who works at the site, those who live in a town where they receive tax benefits because of the location of a nuclear site and who are familiar with it and have never seen an accident—are less concerned. On the other hand, people who live farther away and have no familiarity with the site are much more likely to be afraid of it. The same phenomenon has been observed in studies of chemical weapons sites.

What a reporter in a town in Ohio really needs to know, for example, is what the public in that part of Ohio—not in the nation as a whole—thinks about a nuclear issue. Reporters should exercise caution in using national poll data to speak for specific places that are likely to be affected. Very often it does not. For example, national polls about crime capture the opinions of many people from the suburbs who feel protected against everything and who are not the ones affected by crime. Assuming that national polls really tell the whole story about a specific area likely to be affected by a nuclear concern could produce a serious misunderstanding.

Pitfalls

Although some people blame the media for hyping the "negative" side of an issue, the reality is that people tend to focus on and remember the negative more than the positive, whether the story is about anthrax or nuclear exposures. Furthermore, reporters often cannot immediately obtain information about what material is at issue, how hazardous it is, how much of it has gone into the environment, or how many people have been exposed, and if there has been an exposure, what is in it and where is it heading. So reporters may not be in a position to say that the exposure consists of a small fraction of the exposure that people would normally get from just being outside in Denver, and much less than they would get from a flight from Los Angeles to New York, or how the exposure compares with background in the local area.

That people who feel stressed have trouble "hearing" and "understanding" information that is being provided to them can have major implications for public perception of risk. Mental noise can get in the way of people's understanding information that is being conveyed to them about risk. And when they are stressed, they may begin to distrust the information, worrying that those who are delivering the message are not listening or do not really care. Sandman (2007) has pointed out that sometimes when people are stressed and distrustful, they take a position opposite of what has just been explained. He refers to this as the "seesaw effect" and suggests that as more information is presented, both positions move toward the center and come to a common understanding. Frankly, as noted earlier, we think that reporters have a critical role to play, especially during crises, but that they, like all of us, are challenged to find the reality that exists. Arguably, if they can find it or even frame it, they are more effective at conveying it than others.

References and Other Resources

Agency for Healthcare Research and Quality home page. *www.ahrq.gov*.

Ansolabehere, S. (2007, June). Public attitudes toward America's energy options: Insights for nuclear energy. MIT-NES-TR-08. *mit.edu/canes/pdfs/nes-008.pdf*.

Bisconti Research, Inc. (2005, May 5–9). *U.S. public opinion about nuclear energy.* Report for Nuclear Energy Institute, Washington, DC.

Bronfman, N., & Cifuentes, L. (2003). Risk perception in a developing country: The case of Chile. *Risk Analysis,* 23(6), 171–85.

Centers for Disease Control and Prevention. Radiation Emergencies home page. *www.emergency.cdc.gov/radiation.*

Cha, Y-J. (2000). Risk perception in Korea: A comparison with Japan and the United States. *Journal of Risk Research,* 3(4), 321–32.

Covello, V., & Sandman, P. (2001). Risk communication: Evolution and revolution. In A. Wolbarst (Ed.), *Solutions to an environment in peril.* Baltimore, MD: Johns Hopkins University Press. *www.psandman.com/articles/covello.htm.*

Fischer, G., Morgan, M. G., Fischhoff, B., Nair, I., & Lave, L. (1991). What risks are people concerned about? *Risk Analysis* 11(2), 303–14.

Greenberg, M. (2008). *Are nuclear facilities LULUS? What do United States data tell us?* Edward J. Bloustein School of Planning and Public Policy, Rutgers University and the Consortium for Risk Evaluation with Stakeholder Participation. Paper prepared for symposium honoring Frank Parker, Vanderbilt University, Nashville, TN.

Greenberg, M., Lowrie, K., Burger, J., Powers, C., Gochfeld, M., & Mayer, H. (2007). Nuclear waste and public worries: Public perceptions of the United States' major nuclear weapons legacy sites. *Human Ecology Review,* 14(1).

Leiter, A. (2008). *The perils of a half-built bridge: Risk perception, shifting majorities, and the nuclear power debate. Ecology Law Quarterly,* 35(1), pp. 31–72. *papers.ssrn.com/sol3/papers.cfm?abstract_id=1001109.*

Peter Sandman's Risk Communication Web Site. *www.psandman.com.*

Rosa, E. (2001). Public acceptance of nuclear power: Déjà vu all over again? *Physics and Society,* 30(2), 1–5.

Rosa, E. (2004). *The future acceptability of nuclear power in the United States.* Paris: Institute Français des Relations Internationales.

Sandman, P. (2007, March 22). *Seesaw your way through ambivalence.* CIDRAP Business Source. *www.psandman.com/CIDRAP/CIDRAP8.htm.*

Slovic, P. (1987). Perception of risk. *Science,* 236(4799), 280–85.

Slovic, P., & Weber, E. (2002, April). *Perception of risk posed by extreme events.* Paper prepared for discussion at the conference Risk Management Strategies in an Uncertain World, Palisades, NY. *www.ldeo.columbia.edu/chrr/documents/meetings/roundtable/white_papers/slovic_wp.pdf.*

Sunstein, C. R. (2005). *Laws of fear: Beyond the precautionary principle.* The Seeley Lectures. Cambridge: Cambridge University Press.

Sunstein, C. R. (2006). *Misfearing: A reply.* U Chicago Law & Economics, Olin Working Paper No. 274. Social Science Research Network. *ssrn.com/abstract=880123.*

U.S. Department of Energy home page. *www.energy.gov.*

U.S. Environmental Protection Agency home page. *www.epa.gov.*

U.S. Nuclear Regulatory Commission. Office of Public Affairs home page. *www.nrc.gov/about-nrc/public-affairs.html.*

Wahlberg, A., & Sjoberg, L. (2000). Risk perception and the media. *Journal of Risk Research,* 3(1), 31–50.

Williams, B., Greenberg, M., & Brown, S. (1999). Determinants of perceptions of trust among residents surrounding the Savannah River Site. *Environment and Behavior,* 31(3), 354–71.

Xie, X., Wang, M., & Xu, L. (2003). What risks are Chinese people concerned about? *Risk Analysis, 23*(4), 685–95.

Risk Communication about Nuclear Materials

Written by Karen W. Lowrie, based in part on interviews
with Paul Slovic and Seth Tuler, with review
by Caron Chess and comments by Sandra Quinn

Background

The managers and owners of facilities that process, store, or transport nuclear materials need to provide information to the residents of surrounding communities and all those potentially impacted by risks associated with these activities, and they need to listen to community concerns. Risk communication refers to the content and format of the risk messages that flow between agencies and organizations responsible for nuclear sites and the public, including residents and media outlets in nearby communities. According to the research of many academics and reports of the National Academy of Sciences, risk communication should be a dialogue with communities, not merely the one-way transmission of risk information. The risk communication is often done by public relations or community relations staff that work either for the U.S. Department of Energy (DOE) or for the site contractor. They communicate with the public primarily by providing written information on handouts or through Web sites and by appearing at public meetings to make statements or to address questions and concerns.

Risk communication involves two kinds of information: general or routine information and crisis-mode information, which is provided in response to an event that triggers an emergency, such as an accident, a release of contaminants through a spill, or an explosion. General information typically answers questions like: What is radiation? What do they do at the site that creates risk? What is the status of cleanup? Crisis-mode information should address questions like: What happened? Did the event create additional risk for the surrounding area? How have emergency services responded?

Some facilities have established citizen advisory boards (CABs) that act as formal mechanisms for public input into site decisions and as a conduit for communication between the site and the public. There is a wide range of difference between these various boards in terms of how effective they are at improving communication about risk or building more trust between the site and the general public. Nonetheless, sites that have CABs are likely to use them as the primary delivery mechanism and audience for communication that they feel impacts the public. For example, the largest sites in the nuclear weapons complex, such as Savannah River Site (SC), Oak Ridge (TN), and Hanford (WA), have site-specific advisory boards (SSABs) made up of community stakeholders. These organizations have Web sites with information related to hazards and risks, and they hold regular (monthly or bi-monthly) public meetings.

The DOE and other watchdog groups have assembled various risk communication manuals and guidance documents, but there are no regulatory requirements for risk communication programs. There are notification requirements about certain activities, such as shipments that might pass through communities or building demolitions that might impact neighborhood communities, but these do not have to address risk specifically (10 CFR pt. 71).

Identifying the Issues

Differences often exist between the views of technical people about the safety of nuclear plants or waste storage sites and the views of the public. The job of a good risk communicator is to understand and respect these differences, and to present unbiased, accurate, understandable information about risks. Outside experts, academics, and watchdogs, however, have leveled a good deal of criticism against current government and industry-based risk communication, charging that many of the current written materials and presentations used to explain risks are too complex and not well-designed for the public. Often basic information about, for example, radiation science does not talk about implications for small amounts of exposure. Critics feel that most of the rare primers on radiation and risk that are written in lay language are not very good. Furthermore, they have not been tested well enough to reveal whether risk messages are being transmitted in accurate ways that really improve understanding.

Because there are many influences that affect how people react to risk information and many different circumstances under which risk is discussed, it is difficult to develop general guidance on how to best communicate risk. Experts advise that one of the most important points is to focus on informing and not on persuading. If people think that the site official is trying to persuade them not to worry, they may become more suspicious and distrustful.

For example, it is important that the risk communication does not emphasize point estimates of risk, when in reality there is a level of uncertainty. It is best to admit the uncertainty and also to address the concerns that people care most about. Risk communication is most effective when people have enough facts to help them make informed decisions.

In addition, to be effective agencies must address the interests and concerns of members of the public. For example, agencies should be prepared to explain who made the decisions and why. They might want to raise issues about whether illnesses in the community are related to the risks under discussion. Future land use and economic issues may be concerns. Scientists may not consider risks to quality of life (such as property values), but these are legitimate concerns of affected communities.

Because trust is such an important factor, some site personnel responsible for risk are constantly trying to develop better relations with the public. For the general or routine communication, they may try to provide a greater quantity of information or put more of it on interactive Web sites. Many Web sites also give guided tours of facilities to improve public image and build trust. Active risk communication is often reduced in times that are incident-free, when the public may not be questioning risks because they are accustomed to the facilities that have been there, in most instances, for decades. One of the most active locations where risk communication is playing a role is the controversy surrounding the Yucca Mountain storage facility. Even though most of the technical analyses have been completed, the state of Nevada is still opposing the use of the site. (See "Nuclear Waste Policy.") In this politically charged case, the risk communication presented on the issue is unlikely to markedly change the arguments used by opposing sides.

Some organizations conduct risk communication better than others. One example of a site with an effective process is the Fernald Environmental Management site in Ohio, managed by the DOE. Site officials worked with local citizens to put together a scenario-based model that describes eight different types of people and the typical exposures and risks to them. So when the facility reports about risks, people can determine which of the eight "types" most represents them and interpret the information in a way that makes it relevant to their lives.

Another way to improve risk communication is to create citizen advisory groups or other channels of formal or informal community input. When local citizens help develop the risk communication program, it is likely to be more acceptable to them and therefore more effective. In addition, if a citizens group has helped develop the program, the general public is more likely to trust the information that comes out of it. Finally, most sites have outlined a communication network (that is, a process is in place), but they may not have given much attention to the content and potential effectiveness of the actual message in the event of an emergency. Sites must prepare for crisis-mode communication, addressing the issue before any crisis occurs. And because radiation and risks are difficult phenomena to explain, sites should test their risk communication programs. Although it may appear that crisis-mode communication has improved, its effectiveness is difficult to evaluate because in recent years there have been very few alarming incidents at nuclear sites. More studies are needed to evaluate the content and delivery of risk messages and whether they actually lead to better understanding and acceptance of the information.

What Reporters Need to Know

Risk communication during a crisis should include information about how much hazardous or radioactive material has been released and under what conditions people will be exposed. Reporters covering the story should determine whether the risk communication accurately provides information about the event or circumstance and successfully translates complicated language about substances, measurements, and health effects to a lay audience. Important questions are: What is the hazard? Who might be exposed? At what levels? What is the response?

Reporters should keep in mind that a great deal of scientific study has shown that people do not trust experts because they define risk differently. To an expert, risk is simply a function of the amount of radiation and the probability of expected health effects. (See "Radionuclides.") The public's response, however, is based on a feeling of not having control and awareness that the exposure is involuntary and that it can cause cancer, a particularly dreaded risk. The history of the situation may also have a profound effect on risk communication. (See "Public Perceptions.") Thus, beyond the technical communication is the issue of competence and trust in the people running the agency. People want to know whether someone is to blame and whether the release is something normal. Therefore, additional important questions are: Will people trust the reported levels? How can the public know whom they can trust? Emphasis needs to be placed on reaching out to credible community organizations and leaders to hear their responses and questions about the risk situation.

A reporter should also look at what kinds of comparisons are presented and whether they are valid. Can a lay person really compare the risks as they are presented? For example, if the risk communicator compares background radiation with a level of radiation that is due to site operations, people may perceive them quite differently because these phenomena have different causes: the first is necessary or "natural" and therefore accepted; the other may not seem acceptable because it is not voluntary. Other dimensions that are important in risk comparisons are to describe risk in terms of similar outcomes. It may not be useful to compare a risk of getting cancer from radiation at a site to the chances of getting struck by lightning; it is more useful to compare it to the chances of getting cancer from natural sources. (For a list of questions to ask about risk comparisons, consult National Research Council, 1989, particularly the appendix.) Another important question to consider is whether the communication has looked at all relevant types of risk. For example, people might not be concerned only about a health risk but also about economic risks, such as the effect of a risk on jobs and the local economy.

Misunderstandings

In citing epidemiological studies of long-term health effects, reporters should keep in mind that people often misunderstand the use of the term *statistical power.* A study may not find a positive association or observed effect between exposures and disease and thus report "no effect." That does not mean, however, that there was proof that there was no effect at all. There could still be a possible effect that could not be documented with statistical power. For example, the study may not have had enough cases or it may have had too many uncertainties. A reporter should ask about the real findings of the study, including assumptions and uncertainties, rather than accept a one-sentence summary. In addition, a reporter may want to ask about other studies and also explain for readers how the scientific process works.

Reporters should also be aware that it is not always necessary to present every side of issues. On the contrary, if there is one overriding scientific interpretation of an issue and only a small number of people espouse another view, reporters do not need to force a "debate" to make a story more dramatic. In fact, doing so can create more misunderstandings. For example, it is commonly accepted that risks from certain hazards follow a linear threshold model (see "Radionuclides"), but a minority of people believe the hormesis model, or the idea that some small amount of a contaminant can be good because it stimulates the body to repair itself, and the body adapts to it up to a certain point when it can become bad. Hormesis has been demonstrated for such elements as selenium, iodine, and sunlight. It is important to be wary of this one and other arguments regarding the subjects discussed in this volume, if they are not widely held. When or how to report them is a challenge. A good reference for commonly accepted models and understandings about risk is the series of reports from the National Academy of Sciences called *Biological Effects of Ionizing Radiation* (*BEIR*). (See brief about most recent *BEIR* report at *dels.nas.edu/dels/rpt_briefs/beir_vii_final.pdf*).

Pitfalls

Remember the important principle that risk communicators often use risk comparisons to help people understand esoteric units. For example, they will compare a level of radiation to natural background material or to other types of risk (e.g., this is the same as the risk of having a car accident). These types of comparisons can be dubious because a person might not accept a risk just because it is "lower" than another risk—while the communication may imply that people should accept it. The public may be skeptical or even angry about these comparisons because they do not take into account the various factors that affect risk perception, such as whether it is voluntary or involuntary, what the risk is to future generations, who is profiting, and who is losing.

Reporters should keep in mind that using risk comparisons can also ignore the wide range of uncertainty that is often present in risk calculations. Also, it is important to make clear that there is a difference between relative risk statements ("This might double your risk of cancer") and absolute risk statements ("This might cause a 1 in 1 million risk of cancer"). Good risk communication should include both. A reporter might seek clarification of these concepts from epidemiologists and public health professionals.

It is important to respect the fact that when the public disagrees with a risk assessment, they are not necessarily irrational, but rather they are reacting to a lot of factors that are left out of the communications. Try to evaluate these factors.

References and Other Resources

National Council on Radiation Protection and Measurements home page. *www.ncrp.org*.

National Research Council. (1989). *Improving risk communication*. Washington, DC: National Academies Press. Available at *books.nap.edu/catalog.php?record_id=1189*.

Rocky Mountain Peace and Justice Center home page. *www.rmpjc.org*.

Society for Risk Analysis home page. *www.sra.org*.

U.S. Department of Energy. Environmental Management, Site Specific Advisory Board home page. *www.em.doe.gov/stakepages/ssababout.aspx*.

U.S. Department of Health and Human Services. *Radiation emergencies and event management. www.hhs.gov/disasters/emergency/manmadedisasters/radiation/index.html.*
U.S. Nuclear Regulatory Commission home page. *www.nrc.gov.*
U.S. Nuclear Regulatory Commission. (2004). *Effective risk communication.* NUREG/BR-0308. Washington, DC: U.S. Government Printing Office.
West, B., Lewis, J., Greenberg, M., Sachsman, P., & Rogers, R. (2003). *The reporter's environmental handbook* (3rd ed.). New Brunswick, NJ: Rutgers University Press.

Part III: Additional Resources

History of Nuclear Power in the United States and Worldwide

Written by Karen W. Lowrie, with comments by Thomas Cotton

Background

The concept of the atom has existed for many centuries, going back to Greek philosophers who surmised that all matter is made up of tiny particles (*atomos* is Greek for indivisible). By the turn of the 20 century, physicists such as Ernest Rutherford, called the "father of nuclear science," suspected that the atom, if disintegrated into smaller particles, could release a large amount of energy. Following Rutherford's early experiments with radioactivity, other scientists discovered that radioactive elements have a number of isotopes (different forms of an element having the same number of protons in the nucleus but different numbers of neutrons) and that artificial radionuclides could be produced by bombarding atoms with protons, and more effectively with neutrons.

In 1934, the Italian physicist Enrico Fermi showed that neutrons can split many kinds of atoms. His experiments with splitting uranium resulted in elements much lighter than uranium. In similar experiments, the German scientists Otto Hahn and Fritz Strassman split uranium (atomic number 92) with neutrons to create elements with only about half the atomic mass. Another scientist, Lise Meitner, further proved that the split of the nucleus caused some of the original mass to change to energy, confirming Albert Einstein's theory about the relationship of mass to energy.

A group of scientists from Europe and the United States began to believe a self-sustaining chain reaction might be possible that would create large amounts of energy. Their belief was based on the discovery that fission releases not only energy but also additional neutrons that cause fission in other uranium nuclei. It would happen with enough uranium under proper conditions. The amount of uranium needed to make a self-sustaining chain reaction is called a critical mass. The chain reaction concept was important to the development of atomic bombs. Power stations would have to introduce neutron-absorbing material to control the chain of nuclear reactions, even though uranium used in nuclear power stations does not contain enough of the readily fissionable isotope of uranium to allow a nuclear explosion under any conditions.

By November of 1942, scientists gathered at the University of Chicago began construction of the world's first nuclear reactor, which became known as Chicago Pile-1. The pile, consisting of uranium and cadmium control rods (to absorb neutrons) placed in a stack of graphite, was erected on the floor of a squash court beneath the University of Chicago's athletic stadium. When the control rods were pulled out on December 2, 1942, more neutrons were available to split atoms. The chain reaction sped up and became self-sustaining. Fermi and his team ushered the world into the nuclear age with the first nuclear reactor.

Most of the early atomic research focused on developing an effective weapon for use in World War II, under the code name Manhattan Project. But after the war was over, the U.S. Congress created the Atomic Energy Commission (AEC) to explore peaceful uses of nuclear technology, and researchers worldwide began to focus on ways to harness the tremendous power of nuclear reactions into the generation of electricity. Immediately after the war, reactor research was kept under very strict government control. The AEC oversaw the construction of the first experimental breeder reactor in Idaho (selected in 1949 as the National Reactor Testing Station). A breeder reactor produces both energy and additional fissionable material in the chain reaction. This reactor was completed in 1951 and became the first to produce electricity from nuclear energy. The BORAX III experimental reactor began producing power for Arco, Idaho, on July 17, 1955.

A major goal of nuclear research in the mid-1950s was to show that nuclear energy can produce large amounts of electricity for commercial use. Government emphasized the beneficial uses of the atom and distanced it from the vision of the destructive mushroom cloud. After President Eisenhower's "Atoms for Peace" speech at the United Nations in 1953, calling for greater international partnership in developing nuclear energy, he signed the Atomic Energy Act of 1954, which would spur the expansion of a civilian nuclear power program. The AEC sponsored the world's first large-scale commercial electricity plant powered by nuclear energy, located in Shippingport, Pennsylvania. It became operational in 1957. Private industry became more involved in development, and the nuclear power industry in the United States grew rapidly in the 1960s. Westinghouse and General Electric designed prototypes and began operating the first fully commercial reactors in the early 1960s.

This new form of energy also allowed development of compact long-lasting power sources that could be used to propel ships and submarines. The pressurized water reactor (PWR) was developed for naval use, and the USS *Nautilus,* launched in 1954, became the first nuclear-powered submarine. By the end of the 1950s the United States and Soviet Union also launched nuclear-powered surface vessels. Today over 150 ships, primarily naval vessels, use nuclear power, but Russia and the United States are decommissioning many of their nuclear submarines.

In 1964, President Johnson signed the Private Ownership of Special Nuclear Materials Act, which allows private ownership of uranium fuel by the nuclear power industry. By 1971, 22 commercial nuclear power plants were in full operation in the United States, with 41 more ordered in 1973. A policy change occurred in 1974, when the Energy Reorganization Act divided AEC functions between two new agencies—the Energy Research and Development Administration (ERDA) and the Nuclear Regulatory Commission (NRC), the former to carry out research and development and the latter to regulate nuclear power. ERDA functions were transferred 4 years later to the newly created Department of Energy (DOE).

In the 1970s and 1980s, however, growth of the nuclear industry slowed for a number of reasons, including decreased demand for electricity and growing public concern about environmental issues such as reactor safety and waste disposal. Two events in the late 1970s made it less likely that nuclear power would continue or expand its role as a preferred energy source in the United States and even gave rise to widespread opposition to nuclear power. In 1977, President Jimmy Carter announced that the United States would defer indefinitely plans for reprocessing spent nuclear fuel to recover and recycle usable fissionable material. This announcement closed what had been expected to be the path for disposition of spent fuel and focused attention on waste management as an issue tied to the production of nuclear power.

Further, on March 28, 1979, the worst, and only, major accident in U.S. commercial reactor history occurred at the Three Mile Island nuclear power station near Harrisburg, Pennsylvania. The accident was caused by a loss of coolant from the reactor core due to a combination of mechanical malfunction and human error. Although no one was injured and no radiation overexposures occurred, the NRC quickly reacted by imposing stricter reactor safety regulations. When the Chernobyl accident occurred in 1986 in the former Soviet Union, causing large amounts of radiation to escape from two explosions, exposing millions of people, fears and suspicions were again raised in segments of the general public about the safety and risks of these plants. (See "Three Mile Island and Chernobyl.") In the United States, no new plants have been ordered since 1974 and many reactor orders from the 1970s were canceled.

Austria (1978), Sweden (1980), and Italy (1987) have voted by referendum to oppose or phase out nuclear power.

Another problem with the nuclear industry was that, despite AEC chairman Lewis Strauss's promise of energy that would be "too cheap to meter," the program had been plagued by huge cost-overruns, unexpected outages, heavy regulation, and expensive repairs that resulted in more expensive power than promised. The average nuclear plant completed in 1983 cost $1.7 billion, or about 10 times the cost of the average plant in the early 1970s. (See "The Economics of Nuclear Power.")

Nonetheless, by the early 1990s there were 110 commercial nuclear power plants in the United States producing about 22% of the electricity. The Energy Policy Act of 1992 sought to achieve goals related to safety and design standards and simplified licensing procedures for the next generation of nuclear power plants, and a U.S. nuclear utility consortium began to work on designs for a large advanced reactor. The Energy Policy Act of 2005 contained financial incentives to encourage construction of new nuclear power plants. With the resurgence of interest in nuclear power as a low-carbon electricity source, there are currently 22 licenses applications for new nuclear power plants (33 units) expected to be processed by 2010.

Nuclear Power Worldwide

In the 1950s, the Soviet Union began to refine nuclear reactor designs and develop new ones. By 1954, the Soviets began operating a nuclear-powered electricity generator in the city of Obninsk. It was cooled by water and moderated by graphite, the model that served as a prototype for the Chernobyl plant. The USSR continued to commission more reactors throughout the 1960s and 1970s. Great Britain built a series of reactors, beginning in 1956 with a plant fueled by uranium metal (not enriched uranium as in the U.S. plants) and cooled by gas. Other countries, such as Canada and France, also developed reactors that began operations in the late 1950s and early 1960s. The Canadian reactor design uses unenriched natural uranium moderated by "heavy water" (water with an isotope of hydrogen having an extra neutron). With these exceptions, most countries have ultimately chosen light-water designs (either boiling water or pressurized water) for their nuclear power programs.

By the early 1990s, 31 other countries had a total of 435 nuclear power plants in commercial operation or under construction, with nuclear power accounting for about 17% of the world's electricity in 2007. Still, the United States has twice as many operating nuclear power plants as any other coun-

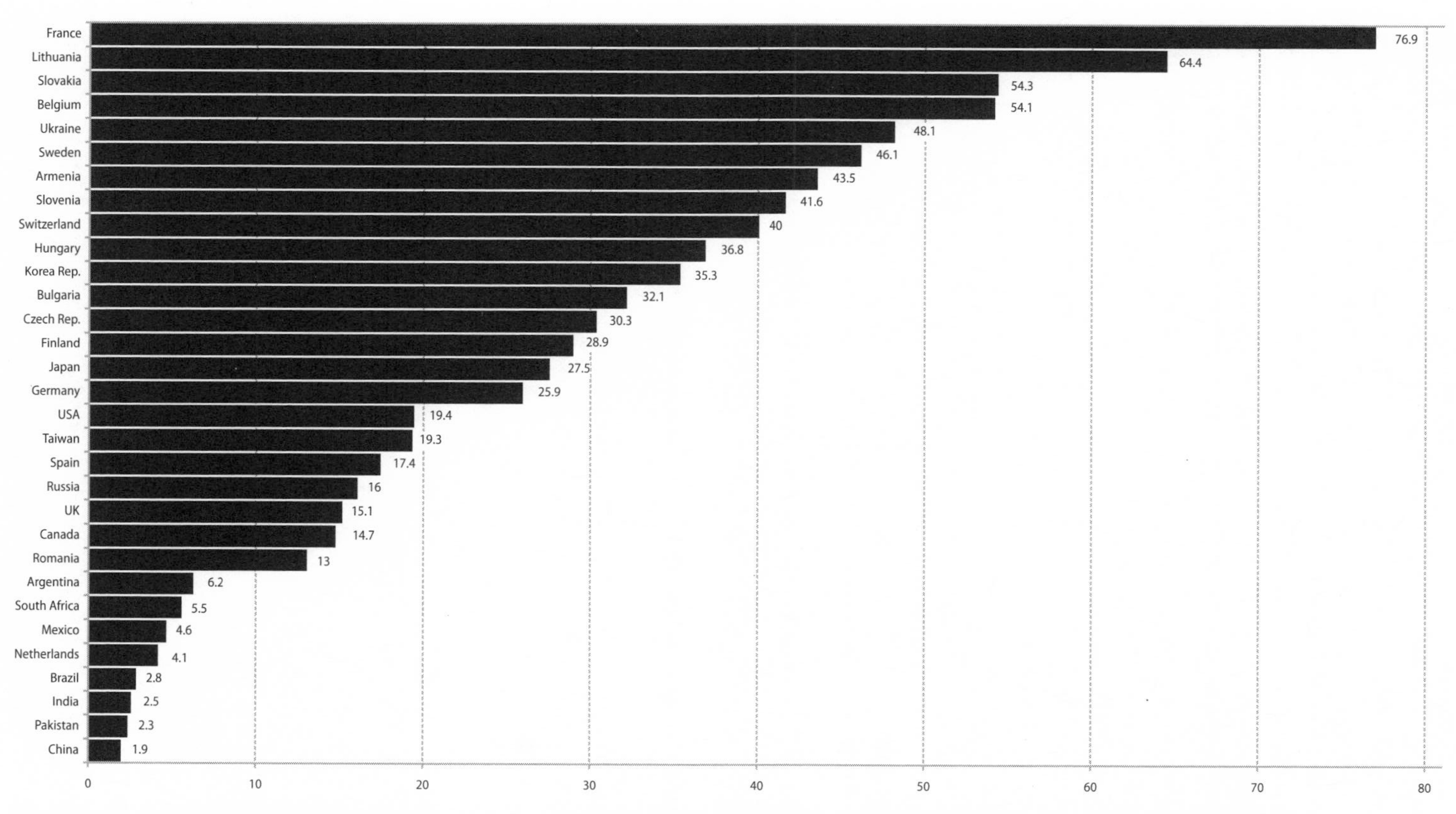

Figure 1. Nuclear share of electricity generation in 2007 (%). Source: IAEA Power Reactor Information System.

try, or about one quarter of the world's operating plants. The United States, France, and Japan together account for over half of the nuclear-generated electricity in the world. Sixteen countries depend on nuclear power for at least a quarter of their electricity. France and Lithuania get more than three quarters of their power from nuclear energy, while Belgium, Bulgaria, Hungary, Slovakia, South Korea, Sweden, Switzerland, Slovenia, and Ukraine get one third or more. Japan, Germany, and Finland get more than a quarter of their power from nuclear energy, while the United States gets almost one fifth (see figure 1).

Selected Resources

American Nuclear Society. (1992). *Controlled nuclear chain reaction: The first 50 years.* La Grange Park, IL: American Nuclear Society.

Glasstone, S. (1979). *Sourcebook on atomic energy* (3rd ed.). Princeton, NJ: Van Nostrand.

Kruschke, E. R., & Jackson, B. M. (1990). *Nuclear energy policy: A reference handbook.* Santa Barbara, CA: ABC-CLIO.

Mazuan, G., & Walker, S. (1984). *Controlling the atom: The beginnings of nuclear regulation, 1946–1962.* Berkeley and Los Angeles: University of California Press.

Simpson, J. W. (1995). *Nuclear power from underseas to outer space.* La Grange Park, IL: American Nuclear Society.

Important Federal Legislation and Regulations

Written by Henry J. Mayer,
with comments by Thomas Cotton

Atomic Energy Act

The Atomic Energy Act (AEA) was originally passed by the U.S. Congress in 1946 following World War II and the demonstration of the power of the atom. The AEA is the fundamental U.S. law on both civilian and military uses of nuclear materials. On the civilian side, the AEA provides for the development and regulation of the uses of nuclear materials and facilities in the United States, declaring the policy that the development and use of atomic energy shall "be directed toward improving the public welfare, increasing the standard of living, strengthening free competition in private enterprise, and promoting world peace." In 1954, the U.S. Congress amended the AEA to encourage the development of commercial nuclear power. These amendments allowed private industry to own and operate nuclear power plants to generate electricity for the public.

The AEA provides authority for setting the standards for the use of nuclear materials to promote the nation's common defense, protect health, and minimize potential danger to life or property. These standards were originally enforced and regulated by the Atomic Energy Commission (AEC). The U.S. Congress created the Nuclear Regulatory Commission (NRC) to replace the AEC when it passed the Energy Reorganization Act of 1974. This act gave the NRC responsibility for regulating various commercial, industrial, academic, and medical uses of nuclear materials and nuclear energy. (Full text at *www.nrc.gov/reading-rm/doc-collections/nuregs/staff/sr0980/rev1/vol-1-sec-1.pdf*, p. 15.)

Nuclear Waste Policy Act

Some forms of nuclear waste (high-level radioactive waste and spent nuclear fuel) remain hazardous for hundreds of thousands of years after production. In the short term, such waste can be stored in facilities with carefully controlled conditions monitored by personnel. However, such monitoring anticipates the continued ability to regulate and operate such facilities. Short-term storage is scattered across the country at numerous sites, thereby exposing large numbers of people to the potential risk associated with nuclear waste. The Nuclear Waste Policy Act (P.L. 97-425) of 1982 recognized the inherent inefficiencies and sustainability issues associated with storage of nuclear waste and established in law that it is the federal government's responsibility to provide facilities for permanent disposal of spent nuclear fuel and high-level radioactive waste from domestic nuclear power plants and from national defense nuclear activities, including production of nuclear weapons and operation of naval nuclear vessels. While previous legislation already provided the U.S. Department of Energy (DOE) with the authority to develop storage facilities and permanent repositories, the Nuclear Waste Policy Act mandated the DOE to proceed expeditiously with siting two permanent disposal facilities (known as geologic repositories) and to license, construct, and operate the first.

The act required that the generators of nuclear waste pay for the construction and operation of permanent disposal facilities. The act established a fund generated by fees charged to nuclear waste generators. The fee was initially set at one tenth of 1 cent per kilowatt-hour but must be adjusted as needed to ensure that the full costs of disposal are borne by the generators of the waste. It authorized the DOE to enter into contracts with the owners and generators of nuclear waste to begin accepting such waste for disposal by January 31, 1998, in exchange for payment of the fees. It established the Office of Civilian Radioactive Waste Management as a program of the DOE, with responsibility for siting, building, and operating the facility, termed a repository. The Environmental Protection Agency (EPA) is responsible for developing standards for protecting the public from exposure to radiation, and the NRC will ultimately authorize DOE to construct and then operate the facility once NRC determines that EPA's standards are met.

A 1987 amendment focused all site investigations on a single site, Yucca Mountain in Nevada. It created the Nuclear Waste Technical Review Board, comprising 10 presidentially appointed scientists, to act as an independent source of expert advice on the scientific and technical aspects of the DOE's handling of the disposal process (Kubiszewski, 2006). The amendment also authorized the DOE to construct a facility for temporary storage of spent fuel, termed "monitored retrievable storage" (MRS). To address the concern that construction of an MRS facility might delay the construction of a permanent repository at Yucca Mountain, the amendment stipulated that construction of the MRS facility may not begin until the NRC authorizes a repository and limits the amount of spent fuel that can be stored at the MRS. It also stipulated that the MRS may not be located in Nevada. (See "Nuclear Waste Policy.")

The Nuclear Waste Policy Act favors the consideration in the siting process of natural geologic features that could safely isolate materials underground and establishes procedures for studying potential sites as well as a timeline for the process. As mandated by the act, the DOE undertook a study of potential sites for permanent nuclear waste disposal, considering nine sites, in Louisiana, Texas, Mississippi, Utah, Nevada, and Washington for the first repository. By 1986, potential sites were narrowed to Yucca Mountain, Nevada, Deaf Smith County, Texas, and Hanford, Washington. Because of rising site assessment costs, Congress passed the Nuclear Waste Policy Amendments Act in 1987, limiting site assessment "characterization" work to Yucca Mountain and directing the DOE to report to Congress between January 1, 2007, and January 1, 2010, on the need for a second repository.

Yucca Mountain was selected for the repository in part because of geologic features that allow the placement of waste deep below the surface in a water-free environment. Before spent nuclear fuel from commercial power plants is shipped by train to Yucca, it would be placed in "transportation, aging, and disposal" containers (TAD containers) and then in radiation-shielding transportation casks. At Yucca, waste would be removed from the casks and cooled off, if necessary, on aging pads before being encased in additional corrosion-resistant disposal containers. The containers would then be deposited, or emplaced, into underground tunnels. The tunnels, also known as drifts, would be about 1,000 feet below the surface, and, on average, 1,000 feet above the water table to minimize natural exposure to moisture.

In 2002, Congress and the President formally approved the Yucca Mountain site as a permanent disposal site, subject to issuance of a license by the NRC. As allowed by the Nuclear Waste Policy Act, the governor of Nevada submitted a "notice of disapproval," but his veto was overridden by a resolution approved by Congress and the President following procedures laid out in the act.

The DOE submitted a license application to the NRC in June 2008, requesting authorization to construct a national repository for the disposal of spent nuclear fuel and high-level radioactive waste at Yucca Mountain. The application consisted of a letter describing its purpose, accompanied by attachments that contain general information and a safety analysis report. The general information portion of the application provides an overview of the repository's engineering design concept and describes the natural features of the site. The safety analysis report (the main technical document in the licensing process) is intended to demonstrate how the repository can be constructed, operated, and closed in a manner that protects public and worker health and safety and preserves the quality of the environment. In September, the NRC announced that it had docketed the DOE's license application. This decision was based on the NRC staff's conclusion that the application is thorough and complete enough for it to begin its formal technical review.

"Final Supplemental Environmental Impact Statement for a Geologic Repository for the Disposal of Spent Nuclear Fuel and High-Level Radioactive Waste at Yucca Mountain, Nye County, Nevada–Nevada Rail Transportation Corridor" (DOE/EIS-0250F-S2), referred to as the final Nevada Rail Corridor SEIS, was released by the DOE in July 2008. In October the DOE announced that it will build a rail line in the proposed Caliente corridor to connect the Yucca Mountain nuclear waste repository with existing rail lines. The line will connect to existing rails near Caliente in Lincoln County, Nevada, head west through Nye County, skirting the Nevada Test and Training Range, and then head south through Esmeralda County to Yucca Mountain. The DOE estimates that it will cost $2.57 billion to construct the line.

Permanent disposal would begin in 2017. However, this schedule could be delayed by funding limitations, problems obtaining permits, or litigation (See the DOE Web site on the Yucca Mountain repository at *www.ocrwm.doe.gov/ym_repository/index.shtml* for studies and articles.) The delay in the repository past the statutory start-up date of January 31, 1998, has already led to court decisions establishing the government's liability for failing to accept spent fuel beginning in 1998 as required by the contracts signed with nuclear utilities under the Nuclear Waste Policy Act. Litigation settlements or damages for that failure are paid out of the Judgment Fund, which is funded by taxpayer dollars, rather than from the Nuclear Waste Fund. The DOE estimates that the potential liability would amount to about $7 billion if Yucca Mountain begins operation in 2017 and that a delay until 2020 could cost an additional $4 billion.

The full text of the Nuclear Waste Policy Act, as amended in 1987, can be found at *www.ocrwm.doe.gov/documents/nwpa/css/nwpa.htm*.

Low-Level Radioactive Waste Policy Act of 1980

At the urging of the National Governors Association, Congress enacted the Low-Level Radioactive Waste Policy Act of 1980, giving each state the responsibility of disposing of low-level radioactive waste generated within its borders. The act leaves it to the state's discretion whether such methods would entail establishing a site within their own territory or joining an interstate compact to develop a regional facility. A primary objective was to reduce the burden on the three states that were accepting all of the nation's low-level wastes: Washington, South Carolina, and Nevada. By 1985, Congress found the 1986 deadline unattainable. In response, it amended the act by extending the deadline to 1993, at which point the three states had the right to refuse incoming low-level radioactive waste not produced within their borders. The amendment also set a series of penalties for states that failed to attain a set level of progress in their development of waste disposal methods. The act also applied to low-level radioactive waste produced by the federal government, making agencies under federal authority responsible for the disposal of their own waste. A summary of the Low-Level Radioactive Waste Act can be found at *www.eoearth.org/article/Low-Level_Radioactive_Waste_Policy_Act_of_1980,_United_States.*

Ten interstate compacts have been formed to date, but only two regional facilities have been built. Eleven states in the Northwest and Rocky Mountain Compacts ship their waste to Richland, Washington, and as of July 1, 2008, only the 3 states in the Atlantic Compact will be permitted to use their regional disposal facility at Barnwell, South Carolina. The remaining 36 states, District of Columbia, and Puerto Rico have no place to dispose of their more radioactive Class B and C categories of low-level waste.

The full text of the Low-Level Radioactive Waste Policy Act can be found at *www.rmllwb.us/documents/federal-low-level-waste-act.pdf.*

Hazardous Materials Transportation Act, as Amended

The Hazardous Materials Transportation Act (P.L. 101-615) was enacted in 1975 to place the regulation of transportation of hazardous waste under the jurisdiction of the secretary of transportation. The act was enacted in recognition of the special hazards associated with the movement of hazardous waste across county, state, and national lines, and of the need for standardized procedures for moving and documenting hazardous wastes. The Department of Transportation, which requires the training of workers dealing with hazardous wastes, classifies materials as hazardous, issues procedures for management of the transfer of the wastes, and establishes criteria for packaging materials and labeling of risks associated with the various types of hazardous waste. Federal rules for handling hazardous waste supersede local, state, and tribal rules, except when expressly waived by the secretary in situations where local rules are stronger then federal rules and do not impede commerce. (See "Transportation of Nuclear Waste.")

Hazardous material is defined by the Hazardous Materials Transportation Act as "a substance or material in a quantity or form which may pose an unreasonable risk to health and safety or property when transported in commerce" (P.L. 101-615, sect. 1802). Hazardous materials can be shipped by water in container barges, in tank containers on tractor trailers, or in small air shipments for time-critical material, for example (U.S. Congress, 1986). Hazardous materials are classified into nine categories based on their chemical and physical properties. According to the classification of the material, the U.S. Department of Transportation (DOT) is responsible for determining the appropriate packaging materials for shipping or transport. Strict guidelines are furnished for proper labeling or marking of packages of hazardous materials offered for transport, and for placarding of transport vehicles. The Hazardous Materials Transportation Act is enforced by use of compliance orders (49 U.S.C. 1808(a)), civil penalties (49 U.S.C. 1809(b)), and injunctive relief (49 U.S.C. 1810).

Because hazardous waste can be transported by road, rail, air, and water, enforcement of the Hazardous Materials Transportation Act is shared by the Federal Highway Administration, the Federal Rail Administration, the Federal Aviation Administration, the Federal Motor Carrier Safety Administration, and the U.S. Coast Guard, along with the DOT. Under the original legislation, the DOT's Research and Special Projects Administration regulated the manufacture and maintenance of shipping containers and equipment. Under current legislation, the Office of Hazardous Material Safety holds this responsibility within the DOT's Pipeline and Hazardous Safety Materials Administration. The Office of Hazardous Material Safety's enforcement responsibility is limited to packaging manufacturers, retesters and reconditioners, and multimodal shippers of hazardous materials.

The shipment of nuclear waste is regulated by the DOT and several agencies. The DOT regulates radioactive materials as a class of hazardous waste. The DOE currently ships defense nuclear waste and weapons for storage, disposal, or use and will send commercial radioactive waste to a repository once it is in operation. The NRC regulates commercial activities of nuclear power plants and certifies shipping containers for spent fuel and other radioactive materials. The Nuclear Waste Policy Act, as amended, requires the DOE to use shipping containers certified by the NRC for shipments of spent fuel and high-level waste under the act.

In 1984 the DOE and the EPA agreed to the Memorandum of Understanding on Responsibilities for Hazardous and Radioactive Mixed Waste Management. In the memorandum, the DOE agrees to comply with the Office of Hazardous Material Safety regulations that require certified containers, proper placarding, and incident notification.

In 1990, Congress enacted the Hazardous Materials Transportation Uniform Safety Act to clarify conflicting local, state, and federal regulations to encourage uniformity among different highway routing regulations, to develop criteria for the issuance of federal permits to motor carriers of hazardous materials, and to regulate the transport of radioactive materials. Also in 1990, the DOT issued new rules that revised regulations regarding hazard communication, classification, and packaging requirements, based on United Nations recommendations, to facilitate international transportation and ensure consistency with international regulations.

The Hazardous Material Transportation Act was reauthorized in 2005 as Title VII of the Safe, Accountable, Flexible, Efficient Transportation Equity Act—A Legacy for Users (SAFETEA-LU), P.L. 109-59, 119 Stat. 1144. SAFETEA-LU doubled the grant money available to communities for hazardous materials (HAZMAT) emergency preparedness and included many other improvements to the HAZMAT authority and process.

The most recent change to the Hazardous Material Transportation Act came in August 2007 when President Bush signed into law "H.R. 1: Implementing Recommendations of the 9/11 Commission Act of 2007." Under the new legislation, railroad carriers are required to consider safety and security factors, including proximity to population centers, when routing hazardous shipments and will submit annual reports of their carrier activity and hazard analyses to the DOT. The law also requires the Department of Homeland Security to screen 100% of cargo on passenger airlines within 3 years. Incoming foreign containers would be scanned for radiation or other hazards. The Department of Homeland Security has not yet issued regulations pertaining to many requirements of the newly enacted legislation.

CERCLA

The Comprehensive Environmental Response, Compensation, and Liability Act (CERCLA), commonly known as Superfund, was enacted by Congress on December 11, 1980, to govern federal responsibilities for the regulation and remediation of some of the nation's most contaminated sites. It established a clean-up fund with taxes paid by chemical and petroleum industries and enabled the federal government, through the EPA, to respond directly to the releases of contamination or potential releases at closed or abandoned sites. Radiological contaminants such as those found at current or former nuclear sites are addressed by CERCLA, but to avoid dual regulation of sites, the EPA defers to the NRC to remediate radiologically contaminated sites at NRC-licensed facilities being closed down (decommissioned). The EPA generally does not list these NRC sites on the National Priorities List (Superfund list).

In 2002, the EPA and the NRC agreed to a formal memorandum of understanding, outlining the process by which the EPA would monitor the NRC's activities and work with the NRC to regulate remediation and decommissioning of nuclear sites. The EPA/NRC memorandum of understanding reiterates the deferral to the NRC in general, but outlines a "consultation" role for the EPA when groundwater contamination is present in levels that exceed limits set by the EPA, when the NRC is contemplating restricting future use of the site ("restricted release") or when soil contaminants exceed levels established by the memorandum of understanding. In those instances, the NRC is to consult with the EPA on clean-up procedures. In addition, the NRC will defer to the EPA regarding nonradiological contaminants.

The full text of the Comprehensive Environmental Response, Compensation, and Liability Act can be found at *www.access.gpo.gov/uscode/title42/chapter103_.html.*

Other Important Legislation and Regulations

Energy Reorganization Act of 1974, as Amended (P.L. 93-438)

Summary: U.S. Nuclear Regulatory Commission. *Our governing legislation. www.nrc.gov/about-nrc/governing-laws.html#era-1974.*

Text: *www.nrc.gov/reading-rm/doc-collections/nuregs/staff/sr0980/ml022200075-vol1.pdf#pagemode=bookmarks&page=213*

Low-Level Radioactive Waste Policy Amendments of 1985 (Title I) (P.L. 99-240)

Summary: U.S. Nuclear Regulatory Commission. *Our governing legislation. www.nrc.gov/about-nrc/governing-laws.html#llrwpaa-1985.*

U.S. Environmental Protection Agency. *Laws we use. www.epa.gov/rpdweb00/laws/laws_sum.html#llrwpa.*

U.S. Environmental Protection Agency. *Radiation protection at EPA: The first 30 years.* Page 56. *www.epa.gov/radiation/docs/402-b-00-001.pdf.*

Text: *www.nrc.gov/reading-rm/doc-collections/nuregs/staff/sr0980/ml022200075-vol1.pdf#pagemode=bookmarks&page=255*

Energy Policy Act of 1992

Summary: U.S. Environmental Protection Agency. *Radiation protection at EPA: The first 30 years.* Page 57. *www.epa.gov/radiation/docs/402-b-00-001.pdf.*

Text: *thomas.loc.gov/cgi-bin/query/z?c102:H.R.776.ENR:*

Waste Isolation Pilot Plant Land Withdrawal Act of 1992, as Amended

Summary: U.S. Environmental Protection Agency. *Laws we use. www.epa.gov/radiation/laws/laws_sum.html#wipplwa.*

Text: *Compilation of selected energy-related legislation: nuclear energy and radioactive waste.* Page 283. *energycommerce.house.gov/107/pubs/nuke.pdf.*

Uranium Mill Tailings Radiation Control Act of 1978 (P.L. 95-604)

Summary: U.S. Nuclear Regulatory Commission. *Our governing legislation. www.nrc.gov/about-nrc/governing-laws.html#umtrca-1978.*

U.S. Environmental Protection Agency. *Laws we use. www.epa.gov/rpdweb00/laws/laws_sum.html#umtrca.*

U.S. Environmental Protection Agency. *Radiation protection at EPA: The first 30 years.* Page 55. *www.epa.gov/radiation/docs/402-b-00-001.pdf.*

Text: *www.nrc.gov/reading-rm/doc-collections/nuregs/staff/sr0980/ml022200075-vol1.pdf#pagemode=bookmarks&page=365*

NRC User Fees (P.L. 101-508)

Summary: *H.R. 5835 Summary as of 10/27/1990: Conference report filed in House. thomas.loc.gov/cgi-bin/bdquery/z?d101:HR05835:@@@D&summ2=m&.*

Text: *thomas.loc.gov/cgi-bin/query/z?c101:H.R.5835:*

Clean Air Act of 1977, as Amended (selected sections) (P.L. 101-549)

Summary: U.S. Environmental Protection Agency. *Radiation protection at EPA: The first 30 years.* Page 54. *www.epa.gov/radiation/docs/402-b-00-001.pdf.*

Text: *www.epa.gov/oar/caa/contents.html*

Federal Water Pollution Control Act of 1972

Summary: U.S. Environmental Protection Agency. *Radiation protection at EPA: The first 30 years.* Page 54. *www.epa.gov/radiation/docs/402-b-00-001.pdf.*

Text: *www.access.gpo.gov/uscode/title33/chapter26_.html*

National Environmental Policy Act of 1969, as Amended (P.L. 91-190)

Summary: U.S. Nuclear Regulatory Commission. *Our governing legislation. www.nrc.gov/about-nrc/governing-laws.html#natl-environ-policy-act.*

Text: *www.nrc.gov/reading-rm/doc-collections/nuregs/staff/sr0980/ml022200111-vol2.pdf#pagemode=bookmarks&page=363*

Resource Conservation and Recovery Act of 1976, as Amended

Summary: U.S. Environmental Protection Agency. *Radiation protection at EPA: The first 30 years.* Page 55. *www.epa.gov/radiation/docs/402-b-00-001.pdf.*
U.S. Environmental Protection Agency. *Laws we use. www.epa.gov/rpdweb00/laws/laws_sum.html#rcra*

Text: *www4.law.cornell.edu/uscode/42/ch82.html*

Comprehensive Environmental Response, Compensation, and Liability Act (CERCLA) of 1980, as Amended

Summary: U.S. Environmental Protection Agency. *Radiation protection at EPA: The first 30 years.* Page 56. *www.epa.gov/radiation/docs/402-b-00-001.pdf.*
Environmental Protection Agency. *Laws we use. www.epa.gov/rpdweb00/laws/laws_sum.html#llrwpa.*

Text: *www4.law.cornell.edu/uscode/42/ch103.html*

Nuclear Nonproliferation Act of 1978 (P.L. 95-242)

Summary: U.S. Nuclear Regulatory Commission. *Our governing legislation. www.nrc.gov/about-nrc/governing-laws.html#nnpa-1978.*

Text: *www.nrc.gov/reading-rm/doc-collections/nuregs/staff/sr0980/ml022200111-vol2.pdf#pagemode=bookmarks&page=376*

International Atomic Energy Agency Participation Act of 1957 (P.L. 85-177) and the Statute of the International Atomic Energy Agency

Summary: *History of the IAEA. www.iaea.org/About/history.html.*

Text: IAEAPA: *www.nrc.gov/reading-rm/doc-collections/nuregs/staff/sr0980/vol2.pdf (page 433)*
IAEA: *www.iaea.org/About/statute_text.html*

International Security Assistance and Arms Export Control Act of 1976 (P.L. 94-329)

Summary: Gerald Ford. *630: Statement on signing the International Security Assistance and Arms Export Control Act of 1976.* July 1, 1976. *www.presidency.ucsb.edu/ws/index.php?pid=6167.*
Richard F. Grimmett. Congressional Research Service. *U.S. defense articles and services supplied to foreign recipients: Restrictions on their use.* March 14, 2005. *www.law.umaryland.edu/marshall/crsreports/crsdocuments/RL3098203142005.pdf.*

Text: Partial text in *Legislation on foreign relations through 2005.* Page 879. *foreignaffairs.house.gov/archives/109/24796.pdf.*

International Security and Development Cooperation Act of 1980 (P.L. 96-533)

Summary: *H.R. 6942 Summary as of 11/12/1980: Conference report filed in House. thomas.loc.gov/cgi-bin/bdquery/z?d096:HR06942:@@@D&summ2=3&.*

Text: Partial text in *Legislation on foreign relations through 2005.* Page 837. *foreignaffairs.house.gov/archives/109/24796.pdf.*

International Security and Development Cooperation Act of 1981 (P.L. 97-113)

Summary: Ronald Reagan. *Statement on signing international security and foreign assistance legislation.* December 29, 1981. John T. Woolley and Gerhard Peters. The American Presidency Project. Santa Barbara, University of California (hosted); Gerhard Peters (database). *www.presidency.ucsb.edu/ws/?pid=43393.*

Text: Partial text in *Legislation on foreign relations through 2005.* Page 819. *foreignaffairs.house.gov/archives/109/24796.pdf.*

Convention on the Physical Protection of Nuclear Material Implementation Act of 1982 (P.L. 97-351)

Summary: Ronald Reagan. *Statement on signing the Convention on the Physical Protection of Nuclear Material Implementation Act of 1982.* October 19, 1982. John T. Woolley and Gerhard Peters. The American Presidency Project. Santa Barbara, University of California (hosted); Gerhard Peters (database). *www.presidency.ucsb.edu/ws/?pid=41882.*

Text: *www.law.cornell.edu/uscode/html/uscode18/usc_sec_18_00000831----000-.html*

Iran-Iraq Arms Nonproliferation Act of 1992 (P.L. 102-484)

Summary: J. Christian Kessler. U.S. Department of State. *United States law and policy: Nonproliferation sanctions. www.exportcontrol.org/library/conferences/1379/Kessler1.pdf.*

Text: *In National Defense Authorization Act for Fiscal Year 1993. thomas.loc.gov/cgi-bin/query/D?c102:7:./temp/~c102XOuliZ::*

Subtitle B-North Korea Threat Reduction of 1999 (P.L.106-113)

Summary: *H.R. 3427 Summary as of 11/17/1999: Introduced. thomas.loc.gov/cgi-bin/bdquery/z?d106:HR03427:@@@D&summ2=m&.*

Text: *frwebgate.access.gpo.gov/cgi-bin/getdoc.cgi?dbname=106_cong_public_laws&docid=f:publ113.106*

Iran Nonproliferation Act of 2000 (P.L. 106-178)

Summary: J. Christian Kessler. U.S. Department of State. *United States law and policy: Nonproliferation sanctions. www.exportcontrol.org/library/conferences/1379/Kessler1.pdf.*

H.R. 1883 Summary as of 2/24/2000: Passed Senate Amended. thomas.loc.gov/cgi-bin/bdquery/z?d106:HR01883:@@@D&summ2=m&.

Text: *frwebgate.access.gpo.gov/cgi-bin/getdoc.cgi?dbname=106_cong_bills&docid=f:h1883enr.txt.pdf*

References and Other Resources

Environmental Protection Agency. (2006, September 22). *Laws we use. www.epa.gov/radiation/laws/laws_sum.html.*

Kubiszewski, I. (2006). Nuclear Waste Policy Act of 1982. In C. L. Cleveland (Ed.), *Encyclopedia of earth.* Washington, DC: Environmental Information Coalition, National Council for Science and the Environment.

U.S. Congress. Office of Technology Assessment. (1986, July). *Transportation of hazardous materials.* OTA-SET-304. Washington, DC: U.S. Government Printing Office.

U.S. Department of Energy. Office of Civilian Radioactive Waste Management Science. *Society and America's nuclear waste: The Nuclear Waste Policy Act, Unit 3; Teacher guide* (2nd ed.). National Information Center, Curriculum Department. Washington, DC: U.S. Government Printing Office.

U.S. Department of Energy. (2007, September 7). *About the project* [Yucca Mountain]. *www.ocrwm.doe.gov/ym_repository/about_project/index.shtml.*

U.S. Department of Energy. (2007, September 7). *Emplacement tunnels. www.ocrwm.doe.gov/ym_repository/studies/engdesign/tunneldesign.shtml.*

U.S. Department of Energy. (2007, September 7). *Transportation, aging, and disposal canister. www.ocrwm.doe.gov/ym_repository/studies/engdesign/wastepackages.shtml.*

U.S. Nuclear Regulatory Commission. (2007, February 15). *Our governing legislation. www.nrc.gov/about-nrc/governing-laws.html#nwpa-1982.*

American Nuclear Society Position Statements

Prepared by Buzz Savage, U.S. Department of Energy

American Nuclear Society position statements represent the Nuclear Society's official position on issues of policy or technical significance. The society's Public Policy Committee is responsible for the adequacy of position statements and submits them to the society's board for approval.

50:	Clearance of Solid Materials from Nuclear Facilities	2008
12:	Fusion Energy	2008
76:	Interim Storage of Used or Spent Nuclear Fuel	2008
82:	Nuclear Power: A Leading Strategy to Reduce Oil Imports	2008
73:	Societal Benefits of Radiation	2008
45:	Nuclear Fuel Recycling	2007
51:	Reactor Safety	2007
21:	Stewardship of Nuclear Engineering Education	2007
81:	The EPA Radiation Standard for Spent-Fuel Storage in a Geological Repository	2006
79:	International Cooperation for Expansion of Nuclear Energy	2006
29:	Maintaining a Viable Nuclear Industry Workforce	2006
24:	Nuclear Facility Safety Standards	2006
44:	Nuclear Power: the Leading Strategy for Reducing Carbon Emissions	2006
78:	The Use of Thorium as Nuclear Fuel	2006
66:	Diversity in the Nuclear Profession	2005
63:	External Costs of Energy Technologies	2005
74:	Fast Reactor Technology: A Path to Long-Term Energy Sustainability	2005
28:	Food Irradiation	2005
33:	Licensing New Nuclear Power Plants	2005
75:	The Need for Deployment of the Next Generation Nuclear Plant Project	2005
56:	The Need for Near-Term Deployment	2005
54:	Price-Anderson Act	2005
72:	The Use of Highly Enriched Uranium for the Production of Medical Isotopes	2005
62:	Use of Nuclear Energy for Desalination	2005

11:	Disposal of Low-Level Radioactive Waste	2004
65:	Realism in the Assessment of Nuclear Technologies	2004
46:	Risk-Informed and Performance-Based Regulations f or Nuclear Power Plants	2004
30:	U.S. Radioisotope Supply	2004
60:	Nuclear Energy for Hydrogen Generation	2003
61:	Nuclear Engineering Licensure	2003
53:	Research and Training Reactors	2003
14:	Use of Nuclear Energy for the Production of Process Heat	2003
47:	The Disposition of Surplus Weapons Plutonium	2002
18:	The Safety of Transporting Radioactive Materials	2002
37:	A Declaration on Sustainable Development by the American Nuclear Society	2001
41:	Health Effects of Low-Level Radiation	2001
55:	Nonproliferation	2001
40:	Space Nuclear Technology	2000
48:	Preserving Experimental Benchmark Work	1999

Background on Key Organizations Related to U.S. Nuclear Programs

Adapted by Michael R. Greenberg, based on text prepared by Buzz Savage, U.S. Department of Energy

Seven key organizations are related to the U.S. nuclear industry: (1) the Institute of Nuclear Power Operations and (2) the Nuclear Energy Institute; on a global scale, (3) the International Atomic Energy Agency and (4) the World Association of Nuclear Operators; and supporting the functions of the U.S. Nuclear Regulatory Commission (NRC) and the U.S. Department of Energy (DOE), (5) the Licensing Support Network, (6) the U.S. Nuclear Waste Technical Review Board, and (7) the Advisory Committee on Nuclear Waste and Materials.

Institute of Nuclear Power Operations (INPO)

www.inpo.info Atlanta, GA

The INPO was formed after the Three Mile Island (TMI-2) event in 1979, based on the recognition that the industry must do a better job of policing itself to ensure that an event of the magnitude of TMI should never happen again. The INPO was charged with establishing standards of excellence against which the plants are measured. An inspection of each member plant is typically performed every 18–24 months. The institute's programs include:

- SEE-IN (an information sharing network)
- EPIX (an equipment failure database)
- National Academy for Nuclear Training
- Events Analysis
- Human Performance
- Accreditation
- Evaluations

INPO maintains a secure private Web site accessible only by the member utilities. INPO serves as the U.S. center for the World Association of Nuclear Operators organization described later.

Nuclear Energy Institute (NEI)

www.nei.org Washington, DC

The NEI serves as an intermediary between the utilities and the NRC on generic nuclear issues. The NEI also serves as a spokesman for the nuclear utilities, conducts public opinion polls, and in general is a conduit for nuclear industry perspectives.

International Atomic Energy Agency (IAEA)

www.iaea.org Vienna, Austria

The IAEA is the United Nations organization that monitors compliance by the member states with nuclear safeguards agreements. The IAEA also promotes nuclear safety on a global scale. IAEA teams conduct OSART (Operational Safety Review Team) inspections at nuclear plants to evaluate nuclear operational safety. The IAEC supports technical needs of member countries.

World Association of Nuclear Operators (WANO)

www.wano.org.uk

WANO was formed after the Chernobyl event to provide a role similar to INPO on a global basis. WANO has regional centers in Atlanta, London, Paris, Tokyo, and Moscow. WANO maintains a secure private Web site accessible only by the member utilities.

Licensing Support Network (LSN)

www.lsnnet.gov

The LSN is a Web-based information system intended to facilitate the discovery process. It supports the NRC's licensing process for a repository at Yucca Mountain, Nevada. All potential parties to the NRC's licensing proceeding, including the DOE and the NRC, place relevant documentary material in the LSN. It can be an important source for journalists because the LSN contains electronically retrievable documentary material relevant to the DOE's license application. The LSN provides the public and potential parties to the NRC's licensing proceeding access to information relevant to the licensing of a repository at Yucca Mountain prior to submittal of the license application. The NRC's regulations for the LSN are found in Title 10, Code of Federal Regulations, part 2, subpart J.

U.S. Nuclear Waste Technical Review Board (NWTRB)

www.nwtrb.gov

The NWTRB is an independent agency of the federal government. Its purpose is to provide independent scientific and technical oversight of the DOE's program for managing and disposing of high-level radioactive waste and spent nuclear fuel.

Advisory Committee on Nuclear Waste and Materials (ACNW&M)

On June 1, 2008, the ACNW&M became the Advisory Committee on Reactor Safeguards. Documents from the former ACNW&M are available on the ACNW&M documents Web page of the NRC Electronic Reading Room Collection (*www.nrc.gov/about-nrc/regulatory/advisory/acnw.html*). For historical information on the ACNW&M, see the ACNW&M History page.

Key Sources

Local Experts

State or local health department staff
University-based nuclear engineering, health physics, public health, and environmental science faculty and staff
Business and not-for-profits

Books

Bodanksy, D. (2004). *Nuclear energy: Principles, practices, and prospects.* New York: Springer.
Garwin, R. L., Charpak, G., & Journe, V. (2002). *Megawatts and megatons: The future of nuclear power and nuclear weapons.* Chicago: University of Chicago Press.
Nero, A. (1979). *A guidebook to nuclear reactors.* Berkeley and Los Angeles: University of California Press.
Waltar, A. (2004). *Radiation and modern life: Fulfilling Marie Curie's dream.* Amherst, NY: Prometheus.

Electronic Sources

American Nuclear Society, *www.ans.org*.
Arms Control Association, *www.armscontrol.org*.
Health Physics Society, *www.hps.org*.
International Atomic Energy Agency, *www.iaea*.org.
Nuclear Energy Institute, *www.nei.org*.
Radwaste.org, *www.radwaste.org*.
Sierra Club's Nuclear Waste Task Force, *www.sierraclub.org/nuclearwaste/*.
Union of Concern Scientists, *www.ucsusa.org*.
U.S. Department of Energy, *www.energy.gov*.
U.S. Department of Transportation, *www.dot.gov*.
U.S. Nuclear Regulatory Commission, *www.nrc.gov*.

Glossary

absolute risk. The proportion of a population expected to get a disease in a specified period. (1)

access hatch. An airtight door system that preserves the pressure integrity of a reactor containment structure while allowing access to personnel and equipment. (2)

accident with release. A transportation accident involving a vehicle (truck or train) in which there is a release of the contents of the container. *See* "Transportation of Nuclear Waste."

accident without release. A transportation accident involving a vehicle (truck or train) that does not involve release of the commodity—similar to the risk for any freight movement on a particular mode, irrespective of the type of freight being moved, because the commodity itself does not cause any harm. *See* "Transportation of Nuclear Waste."

actinides. Any of the series of 15 chemical elements with increasing atomic numbers, including plutonium and uranium, that begins with actinium or thorium and ends with lawrencium on the periodic table.

activation. The process of making a radioactive isotope by bombarding a stable element with neutrons or protons. (2)

active fuel length. The end-to-end dimension of fuel material within a fuel element. (2)

activity. The rate of disintegration (transformation) or decay of radioactive material per unit time. The units of activity are the curie (Ci) and the becquerel (Bq). (2)

ADAMS. *See* Agencywide Documents Access and Management System.

AEC. *See* Atomic Energy Commission.

Agencywide Documents Access and Management System (ADAMS). The system used by the U.S. Nuclear Regulatory Commission (NRC) since 2000 to make information available for public review. It includes documents related to licensing decisions (e.g., the license application), as well as subsequent changes, correspondence between the licensee and the NRC, and inspections reports.

agreement state. A state that has signed an agreement with the U.S. Nuclear Regulatory Commission under which the state regulates the use of byproduct, source, and small quantities of special nuclear material in that state. (2)

air sampling. The collection of samples of air to measure the radioactivity or to detect the presence of radioactive material, particulate matter, or chemical pollutants in the air. (2)

airborne radioactivity area. A room, enclosure, or area in which airborne radioactive materials, made up wholly or partly of licensed material, existing in concentrations that (*a*) exceed the derived air concentration limits or (*b*) would result in an individual present in the area without respiratory protection exceeding, during those hours, 0.6% of the annual limit on intake or 12 derived air concentration-hours. (2)

ALARA (as low as [is] reasonably achievable). Refers to the goal of efforts to minimize radiation exposure as far below the maximum dose limit as is possible and practical.

ALI. *See* annual limit on intake.

alluvium. Sedimentary material (clay, silt, sand, gravel, or similar material), deposited by running water.

alpha particle. A positively charged particle with low penetrating power and a short range that spontaneously emits from some radioactive elements. Alpha particles will generally fail to penetrate the outer layers of skin but can be harmful when ingested.

anion. A negatively charged ion. (2)

annual limit on intake (ALI). The derived limit for the amount of radioactive material that should be taken into the body of an adult worker by inhalation or ingestion in a year. ALI is the smaller value of intake of a given radionuclide in a year by the reference man that would result in a committed effective dose equivalent of 5 rems (0.05 sievert) or a committed dose equivalent of 50 rems (0.5 sievert) to any individual organ or tissue. (2)

anticipated transient without scram (ATWS). One of the "worst case" accidents, consideration of which frequently motivates the U.S. Nuclear Regulatory Commission to take regulatory action. The accident could happen if the system that provides a highly reliable means of shutting down the reactor (scram system) fails to work during a reactor event (anticipated transient). The types of events considered are those used for designing the plant. (2)

APLHGR. *See* average planar linear heat generation rate.

atom. The smallest particle of an element that cannot be divided or broken up by chemical means. It consists of a central core of protons and neutrons, called the nucleus, and electrons that revolve around the nucleus. (2)

atomic energy. Energy released in nuclear reactions. Of particular interest is the energy released when a neutron initiates the breaking up of an atom's nucleus into smaller pieces (fission) or when two nuclei are joined together under millions of degrees of heat (fusion). It is more correctly called nuclear energy. (2)

Atomic Energy Commission (AEC). Federal agency created in 1946 to manage the development, use, and control of nuclear energy for military and civilian applications. Abolished by the Energy Reorganization Act in 1974 and succeeded by the Energy Research and Development Administration (now part of the U.S. Department of Energy) and the U.S. Nuclear Regulatory Commission. (2)

atomic number. The number of positively charged protons in the nucleus of an atom. (2)

attenuation. The process by which the number of particles or photons entering a body of matter is reduced by absorption and scattered radiation. (2)

ATWS. *See* anticipated transient without scram.

auxiliary building. Building at a nuclear power plant, frequently adjacent to the reactor containment structure that houses most of the reactor auxiliary and safety systems, such as radioactive waste systems, chemical and volume control systems, and emergency cooling water systems. (2)

auxiliary feedwater. Backup water supply used during nuclear plant startup and shutdown to supply water to the steam generators during accident conditions for removing decay heat from the reactor. (2)

average planar linear heat generation rate (APLHGR). The average value of the linear heat generation rate of all the control rods at any given horizontal plane along a fuel bundle. (2)

background radiation. Radiation from cosmic sources; naturally occurring radioactive materials, including radon and global fallout as it exists in the environment from the testing of nuclear explosive devices. The typically quoted average individual exposure from background radiation is 360 millirems per year. (2)

Bayesian estimation. A mathematical formulation that predicts the likelihood of an event can be estimated taking explicit consideration of certain contextual features (such as amount of data, nature of decision, etc.). (2)

Bayesian prior. A way to express the context of a Bayesian estimation in which initial data are updated as new data become available. (2)

becquerel (Bq). The international system (SI) unit of radioactivity, replacing the curie, which equals 1 disintegration per second. 37 billion (3.7×10^{10}) becquerels = 1 curie (Ci).

BELLE (biological effects of low-level exposures). The biological effects of low-level exposures to chemical agents and radioactivity.

beryllium. A highly toxic steel-gray metal, which can be used in nuclear reactors as a moderator, reflector, or cladding material. (2)

beta particle. A high energy, high speed particle that spontaneously emits from a nucleus during radioactive decay. A negatively charged beta particle is identical to an electron and positively charged beta particle is called a positron. Large amounts of beta radiation may cause skin burns, and beta emitters are harmful if they enter the body.

beyond design-basis accidents. Accident sequences that are possible but were not originally fully considered in the design process because they were judged to be too unlikely. (2)

binding energy. The minimum energy required to separate a nucleus into its component neutrons and protons. (2)

bioassay. The determination of types, concentrations, and locations of radioactive material in the human body, whether by direct measurement or by analysis of materials excreted or removed from the human body.

biological half-life. The time required for a biological system, such as that of a human, to eliminate, by natural processes, half of the amount of a substance (such as a radioactive material) that has entered it. (2)

biological shield. Absorbing material, such as graphite or concrete, placed around a reactor or radioactive source to reduce the radiation to a level safe for humans.

biomarker. Substances in body tissue or fluids that indicate exposure, effect, and susceptibility. *See* "Impact of Radionuclides and Nuclear Waste Management."

biomass. Organic nonfossil material of biological origin constituting a renewable energy source.

biomass gas. A medium Btu gas containing methane and carbon dioxide, resulting from the action of microorganisms on organic materials such as a landfill. (3)

biome. Entire community of living organisms in a single major ecological area. (4)

boiling water reactor (BWR). A reactor in which water, used as both coolant and moderator, is allowed to boil in the core. The resulting steam can be used directly to drive a turbine and electrical generator, thereby producing electricity. (2)

bone seeker. A radioisotope that tends to accumulate in the bones when it is introduced into the body. An example is strontium-90, which behaves chemically like calcium. (2)

breeder. A reactor that produces more nuclear fuel than it consumes. A fertile material, such as uranium-238, when bombarded by neutrons, is transformed into a fissile material, such as plutonium-239, which can be used as fuel. (2)

British thermal unit (Btu). The amount of heat required to change the temperature of one pound of water one degree Fahrenheit at sea level. (2)

byproduct. Material made radioactive or created during the process of producing or using special nuclear material, or the wastes produced during uranium or thorium mining (tailings).

calibration. The adjustment, as necessary, of a measuring device such that it responds within the required range and accuracy to known values of input. (2)

capability. The maximum load that a generating station can carry under specified conditions for a given period without exceeding approved limits of temperature and stress. (2)

capacity factor (gross). The ratio of the gross electricity generated, for the time considered, to the energy that could have been generated at continuous full-power operation during the same period. (2)

capacity factor (net). The ratio of the net electricity generated, for the time considered, to the energy that could have been generated at continuous full-power operation during the same period. (2)

cask. A heavily shielded container used to store and/or ship radioactive materials. Lead and steel are common materials used in the manufacture of casks. (2)

cation. A positively charged ion. (2)

centrifuge. A machine that uses centrifugal force to separate substances of different densities, to remove moisture or to simulate gravitational effects. Centrifuges can be used to separate the heavier uranium-235 isotope from uranium-238.

cesium 137. A radioactive isotope of cesium used in radiation therapy.

CFR (Code of Federal Regulations). A codification of the general and permanent rules and regulations published in the *Federal Register* by the executive departments and agencies of the U.S. federal government.

chain reaction. A reaction that initiates its own repetition. A fission chain reaction is self-sustaining when the number of neutrons released in a given time equals or exceeds the number of neutrons lost by absorption in nonfissionable material or by escape from the system. (2)

charged particle. An ion; an elementary particle carrying a positive or negative electrical charge. (2)

chemical recombination. Transformation of chemicals during exposure to radiation, in which positively and negatively charged ions may realign themselves into a different chemical configuration.

cladding. The thin-walled metal tube that forms the outer jacket of a nuclear fuel rod. It prevents coolant from corroding the fuel, releasing fission products into the coolant. Aluminum, stainless steel, and zirconium alloys are common cladding materials. (2)

cleanup system. A system used for continuously filtering and demineralizing a reactor coolant system to reduce contamination levels and to minimize corrosion. (2)

coastdown. An action that permits the reactor power level to decrease gradually as the fuel in the core is depleted. (2)

cold pasteurization. Another term for food irradiation, in which food is exposed to ionizing radiation to kill bacteria.

cold shutdown. The term used to define a reactor coolant system at atmospheric pressure and at a temperature below 200°F following a reactor cooldown. (2)

collective dose. The sum of the individual doses received in a given period by a population from exposure to a specified source of radiation. (2)

combined cycle. An electricity-generating technology in which electricity is produced from otherwise lost waste heat exiting from one or more gas (combustion) turbines. The exiting heat is routed to a conventional boiler or to a heat recovery steam generator for use by a steam turbine in the production of electricity. This process increases the efficiency of the electric generating unit. (3)

Commission on Risk Assessment and Risk Management. A commission mandated by the Clean Air Act Amendments of 1990 and charged with making a full investigation of the policy implications and appropriate uses of risk assessment and risk management in regulatory programs under various federal laws to prevent cancer and other chronic human health effects that may result from exposure to hazardous substances. It was disbanded in 1997.

committed dose equivalent (CDE). The dose to some specific organ or tissue that is received from an intake of radioactive material by an individual during the 50 years following the intake. (2)

committed effective dose equivalent (CEDE). The committed dose equivalent for a given organ multiplied by a weighting factor (see 10 CFR 20.1003). (2)

compact. An agreement between two or more states to dispose of low-level radioactive waste regionally.

comparative risk analysis. An environmental decision-making tool used to systematically measure, compare, and rank environmental problems or issue areas that focuses usually on the risks a problem poses to human health, the natural environment and quality of life, and results in a list of issue areas ranked by their relative risks. (5)

composite adversary force. An adversary force used in force-on-force exercises that is trained to standards issued by the U.S. Nuclear Regulatory Commission.

compound. A chemical combination of two or more elements combined in a fixed and definite proportion. For example, water (H_2O) is a compound made up of hydrogen and oxygen.

condensate. Water that has been produced by the cooling of steam in a condenser. (2)

condenser. A large heat exchanger designed to cool exhaust steam from a turbine below the boiling point so that it can be returned as water to be reheated. The heat removed by the condenser is transferred to a circulating water system and is exhausted to the environment, either through a cooling tower or directly into a body of water. (2)

construction recapture. The maximum number of years that could be added to the license expiration date to recover the period from the construction permit to the date when the operating license was granted. A licensee is required to submit an application for such a change. (2)

contact handled (CH) waste. Transuranic waste that has a measured radiation dose rate at the container surface of 200 millirems per hour or less and can be safely handled without special equipment when placed in containers. (6)

containment structure. A gas-tight shell or other enclosure around a nuclear reactor to confine fission products that otherwise might be released to the atmosphere in the event of an accident. (2)

contamination. Undesired radioactive material that is deposited on the surface of or inside structures, areas, objects, or people. (2)

control rod. A rod, plate, or tube containing a material that absorbs neutrons that is used to control the fission activity in a nuclear reactor.

control room. The area in a nuclear power plant from which most of the plant power production and emergency safety equipment can be operated by remote control. (2)

controlled area. At a nuclear facility, an area outside a restricted area but within the site boundary, access to which the licensee can limit for any reason. (2)

coolant. A substance circulated through a nuclear reactor to remove or transfer heat. The most commonly used coolant in the United States is water. Other coolants include heavy water, air, carbon dioxide, helium, liquid sodium, and a sodium-potassium alloy. (2)

cooldown. The gradual decline in reactor fuel rod temperature caused by the removal of heat from the reactor coolant system after the reactor has been shutdown. (2)

cooling tower. A heat exchanger that helps to cool water that was used to cool exhaust steam exiting the turbines of a power plant. Cooling towers transfer exhaust heat into the air rather than into a body of water. (2)

core. The central portion of a nuclear reactor containing the fuel elements, moderator, control rods, and support structures. (2)

core damage frequency. An expression of the likelihood that, because of the way a reactor is designed and operated, an accident could cause the fuel in the reactor to be damaged. (2)

core melt accident. An event or sequence of events that results in the melting of part of the fuel in the reactor core. (2)

cosmic radiation. Penetrating radiation, both particulate and electromagnetic, originating in outer space. Secondary cosmic rays, formed by interactions in the earth's atmosphere, account for about 45–50 millirem of the 360 millirem background radiation that an average individual receives in a year.

counter. A general designation applied to radiation detection instruments or survey meters that detect and measure radiation. The signal that announces an ionization event is called a count.

critical mass. The smallest mass of fissionable material that will support a self-sustaining chain reaction.

critical organ. That part of the body that is most susceptible to radiation damage under the specific conditions under consideration. (2)

criticality. A term used in reactor physics to describe the state when the number of neutrons released by fission is exactly balanced by the neutrons being absorbed (by the fuel and poisons) and escaping the reactor core. A reactor is said to be "critical" when it achieves a self-sustaining nuclear chain reaction, as when the reactor is operating. (2)

crud. A colloquial term for corrosion and wear products (rust particles, etc.) that become radioactive when exposed to radiation. Crud originally stood for the Chalk River Unidentified Deposits at the Canadian nuclear plant Chalk River.

CT scan (computed axial tomography). A method that uses x-rays to create cross-sectional images of the body.

cumulative dose. The total dose resulting from repeated exposures of radiation to the same portion of the body, or to the whole body, over time. (2)

curie (Ci). The basic unit used to describe the intensity of radioactivity in a sample of material. The curie is equal to 37 billion (3.7×10^{10}) disintegrations per second, which is approximately the activity of 1 gram of radium. (2)

daughter products. Isotopes that are formed by the radioactive decay of some other isotope. Radium-226, for example, has 10 successive daughter products, ending in the stable isotope lead-206. (2)

decay, radioactive. Decrease in the amount of any radioactive material with the passage of time due to the spontaneous emission of radiation from an atomic nucleus.

decay heat. The heat produced by the decay of radioactive fission products after a reactor has been shut down. (2)

declared pregnant woman. A woman who is an occupational radiation worker and has voluntarily informed her employer, in writing, of her pregnancy and the estimated date of conception (see 10 CFR 20.1003 and 20.1208). (2)

decommissioning. The process of closing down a facility followed by reducing residual radioactivity to a level that permits the release of the property for unrestricted use. (2)

DECON. A method of decommissioning in which the equipment, structures, and portions of a facility and site containing radioactive contaminants are removed and safety buried in a low-level radioactive waste landfill or decontaminated to a level that permits the property to be released for unrestricted use shortly after cessation of operations. (2)

decontamination. The reduction or removal of contaminating radioactive material from a structure, area, object, or person. Decontamination may be accomplished by (*a*) removing or treating surface material, (*b*) allowing the material stand to decay naturally, or (*c*) covering the contamination to shield the radiation. (2)

deep-dose equivalent (DDE). The external whole-body exposure dose equivalent at a tissue depth of 1 cm (1,000 mg/cm^2). (2)

defense-in-depth. A design and operational philosophy with regard to nuclear facilities that calls for multiple layers of protection to prevent accidents. It includes the use of controls, multiple physical barriers to prevent release of radiation, redundant and diverse key safety functions, and emergency response measures. (2)

departure from nuclear boiling ratio (DNBR). The ratio of the heat flux to cause departure from nucleate boiling to the actual local heat flux or a fuel rod. (2)

departure from nucleate boiling (DNB). The point at which the heat transfer from a fuel rod rapidly decreases because of the insulating effect of a steam blanket that forms on the rod surface when the temperature continues to increase. (2)

depleted uranium. Uranium having a percentage of uranium-235 smaller than the 0.7% found in natural uranium. It is obtained from spent (used) fuel elements or as byproduct tails, or residues, from uranium isotope separation. (2)

derived air concentration (DAC). The concentration of radioactive material in air and the time of exposure to that radionuclide in hours. A U.S. Nuclear Regulatory Commission licensee may take 2,000 hours to represent 1 ALI, equivalent to a committed effective dose equivalent of 5 rems (0.05 sievert). (2)

derived air concentration hour (DAC-hour). The product of the concentration of radioactive material in air (expressed as a fraction or multiple of the derived air concentration for each radionuclide) and the time of exposure to that radionuclide, in hours. A U.S. Nuclear Regulatory Commission licensee may take 2,000 DAC-hours to represent 1 ALI, equivalent to a committed effective dose equivalent of 5 rems (0.05 sievert). (2)

desalination. (*a*) Removal of salts from ocean or brackish water by using various technologies. (*b*) Removal of salts from soil by artificial means, usually leaching. (4)

design-basis accident. A postulated accident that a nuclear facility must be designed and built to withstand without loss to the systems, structures, and components necessary to ensure public health and safety. (2)

design-basis phenomena. Earthquakes, tornadoes, hurricanes, floods, and other natural disasters that a nuclear facility must be designed and built to withstand without loss to the systems, structures, and components necessary to ensure public health and safety. (2)

design-basis threat (DBT). A profile of the type, composition, and capabilities of an adversary. The U.S. Nuclear Regulatory Commission and its licensees use the DBT as a basis for designing safeguards systems to protect against acts of radiological sabotage and to prevent the theft of special nuclear material. This term is applied to clearly identify for a licensee the expected capability of its facility to withstand a threat. (2)

detector. A material or device that is sensitive to radiation and can produce a measurable response signal.

deterministic (probabilistic). Consistent with the principles of "determinism," which hold that specific causes completely and certainly determine effects of all sorts. As applied in nuclear technology, it generally deals with evaluating the safety of a nuclear power plant in terms of the consequences of a predetermined bounding subset of accident sequences.

The term *probabilistic* is associated with an evaluation that accounts for the likelihood and consequences of possible accident sequences in an integrated fashion. (2)

deterministic effect. The health effects of radiation, the severity of which varies with doses and for which a threshold is believed to exist. Radiation-induced cataract formation is an example of a deterministic effect (also called nonstochastic effect). (2)

deuterium. A stable isotope of hydrogen with one proton and one neutron in the nucleus, unlike the more common form that consists solely of a proton.

deuteron. The nucleus of deuterium, containing one proton and one neutron. (2)

differential pressure (dp or dP). The difference in pressure between two points of a system, such as between the inlet and outlet of a pump. (2)

dirty bomb. *See* Radiological dispersal device.

Doppler coefficient. The change in reactivity per degree change in temperature, also referred to as the fuel temperature coefficient of reactivity.

dose. The absorbed dose, given in rads (or in SI units, grays), that represents the energy absorbed from the radiation in one gram of any material. Furthermore, the biological dose or dose equivalent, given in rem or sieverts, is a measure of the biological damage to living tissue from radiation exposure. (2)

dose, absorbed. The amount of energy deposited in any substance by ionizing radiation per unit mass of the substance. It is expressed numerically in rads or grays. (2)

dose equivalent. The absorbed dose in tissue, multiplied by a quality factor and then sometimes multiplied by other necessary modifying factors at the location of interest. (2)

dose rate. The ionizing radiation dose delivered per unit time. For example, rem or sieverts per hour. (2)

dosimeter. A small portable instrument (such as a film badge or pocket dosimeter) for measuring and recording the total accumulated personal dose of ionizing radiation. (2)

dosimetry. The theory and application of the principles and techniques involved in the measurement and recording of ionizing radiation doses. (2)

drywell. The containment structure enclosing a boiling water reactor vessel and its recirculation system. The drywell provides both a pressure suppression system and a fission product barrier under accident conditions. (2)

earthquake, operating basis. An earthquake magnitude for which the power plant has been designed to withstand without undue risk to public health and safety. *See* Design-basis phenomena.

ecological risk assessment. The application of a formal framework, analytical process, or model to estimate the effects of human action or actions on a natural resource and to interpret the significance of those effects in the light of the uncertainties identified in each component of the assessment process. Such analysis includes initial hazard identification, exposure and dose-response assessments, and risk characterization. (4)

effective dose equivalent. The sum of the products of the dose equivalent to the organ or tissue and the weighting factors applicable to each of the body organs or tissues that are irradiated. (2)

effective half-life. The time required for the amount of a radioactive element deposited in a living organism to be diminished 50% as a result of the combined action of radioactive decay and biological elimination. (2)

efficiency, plant. The percentage of the total energy content of a power plant's fuel that is converted into electricity. The remaining energy is lost to the environment as heat. (2)

EIS. *See* environmental impact statement.

electrical generator. An electromagnetic device that converts mechanical energy into electrical energy. Most large electrical generators are driven by steam or water turbine systems. (2)

electromagnetic pulse. A sharp pulse of intense electromagnetic radiation produced by a nuclear explosion. The intense electric and magnetic fields can damage unprotected electrical and electronic equipment over a large area, disrupting communication and power transmission systems. (2)

electromagnetic radiation. A traveling wave motion resulting from changing electric or magnetic fields. Familiar electromagnetic radiation range from x-rays of short wavelength, through the ultraviolet, visible, and infrared regions, to radar and radio waves of relatively long wavelength. (2)

electron. A negatively charged particle that surrounds the nucleus and has a mass 1/1837 that of the proton.

element. One of the 103 known chemical substances, such as hydrogen, nitrogen, or gold, that cannot be broken down further without changing its chemical properties.

emergency classifications. Response by an offsite organization is required to protect local citizens near the site. A request for assistance from offsite emergency response organizations may be required. (2)

emergency core cooling systems (ECCS). A system of pumps, valves, heat exchangers, tanks, and piping that is designed to remove heat from the reactor fuel rods should the normal core cooling system fail.

emergency feedwater. Backup feedwater supply used during nuclear plant startup and shutdown; also known as auxiliary feedwater. (7)

emergency planning zone (EPZ). To facilitate a preplanned strategy for protective actions during an emergency, there are two emergency planning zones (EPZs) around each nuclear power plant—plume exposure pathway EPZ and ingestion exposure pathway EPZ. The exact size and shape of each EPZ is a result of detailed planning that includes consideration of the specific conditions at each site, unique geographical features of the area, and demographic information. This preplanned strategy for an EPZ provides a substantial basis to support activity beyond the planning zone in the extremely unlikely event it would be needed. (8)

encapsulation. The treatment of asbestos-containing material with a liquid that covers the surface with a protective coating or embeds fibers in an adhesive matrix to prevent their release into the air. (4)

engineered controls. Method of managing environmental and health risks by placing a barrier between the contamination and the rest of the site, thus limiting exposure pathways. (4)

enrichment. The process of increasing the concentration of one isotope of a given element. (9)

entomb. A method of decommissioning in which radioactive contaminants are encased in a structurally long-lived material, such as concrete. The entombment structure is appropriately maintained and continued surveillance is carried out until the radioactivity decays to a level permitting decommissioning and ultimate unrestricted release of the property. (2)

environmental impact statement (EIS). A report that documents the information required to evaluate the environmental impact of a project. It informs decision makers and the public of the reasonable alternatives that would avoid or minimize adverse impacts or enhance the quality of the environment. (3)

environmental qualification. A process for ensuring that equipment will be capable of withstanding environmental conditions that could exist when the equipment is required to operate under accident conditions.

erg. A unit of energy.

event notification (EN) system. An internal U.S. Nuclear Regulatory Commission (NRC) system used to track notifications of significant material events that have affected or may affect public health or safety. Significant material events are reported to the NRC Operations Center by NRC licensees, staff of the agreement states, other federal agencies, and the public. (2)

exclusion area. The area surrounding the reactor where the reactor licensee has the authority to determine all activities, including exclusion or removal of personnel and property. (2)

excursion. A sudden, very rapid rise in the power level of a reactor caused by supercriticality. Excursions are usually quickly suppressed by the negative temperature coefficient, the fuel temperature coefficient or the void coefficient (depending upon reactor design), or by rapid insertion of control rods. (2)

exposure. The complex events by which a receptor (human, wildlife, plant) comes into the presence of or contact with a foreign substance (toxic contaminant or essential nutrient) in the environmental media (air, water, soil, or food), and how the substance enters the body through inhalation, ingestion, or dermal uptake.

external radiation. Exposure to ionizing radiation when the radiation source is located outside the body. (2)

externalities. Benefits or costs, generated as a byproduct of an economic activity, that do not accrue to the parties involved in the activity. Environmental externalities are benefits or costs that manifest themselves through changes in the physical or biological environment. (3)

extremities. The hands, forearms, elbows, feet, knees, leg below the knees, and ankles. (Permissible radiation exposures in these regions are generally greater than in the whole body because they contain fewer blood-forming organs and have smaller volumes for energy absorption.)(2)

fallout. The descent of particles contaminated with radioactive material from a radioactive cloud. Fallout is both early (local), reaching the earth within the first day after the explosion, and delayed (worldwide). Delayed fallout can be brought to earth by rain or snow over months and years. (9)

fast fission. Fission of a heavy atom (such as uranium-238) when it absorbs a high energy (fast) neutron. Most fissionable materials need thermal (slow) neutrons in order to fission. (2)

fast neutron. A neutron with kinetic energy greater than its surroundings when released during fission. (2)

feedwater. Water supplied to the reactor pressure vessel or the steam generator that removes heat from the reactor fuel rods by boiling and becoming steam. The steam becomes the driving force for the plant turbine generator. (2)

fertile material. A material, which is not itself fissile (fissionable by thermal neutrons), that can be converted into a fissile material by irradiation in a reactor. There are two basic fertile materials: uranium-238 and thorium-232. When these fertile materials capture neutrons, they are converted into fissile plutonium-239 and uranium-233, respectively. (2)

film badge. Photographic film used for measurement of ionizing radiation exposure for personnel monitoring purposes. The film badge may contain two or three films of differing sensitivities, and it may also contain a filter that shields part of the film from certain types of radiation. (2)

fireball. The sphere of hot gases that form on a nuclear explosion.
fissile material. Any material fissionable by thermal (slow) neutrons. The three primary fissile materials are uranium-233, uranium-235, and plutonium-239.
fission (fissioning). The splitting of a nucleus into at least two other nuclei and the release of a relatively large amount of energy. Two or three neutrons are usually released during this type of transformation. Fission occurs naturally or when an atom's nucleus is bombarded by neutrons. (2)
fission gases. Those fission products that exist in a gaseous state. In nuclear power reactors, they includes primarily the noble gases, such as krypton and xenon. (2)
fission products. The nuclei (fission fragments) formed by the fission of heavy elements, as well as the nuclide formed by the fission fragments' radioactive decay. (2)
fissionable material. Commonly used as a synonym for fissile material, the meaning of this term has been extended to include material that can be fissioned by fast neutrons, such as uranium-238. (2)
flash burn. A burn caused by excessive exposure (of bare skin) to thermal radiation. (3)
flux. The amount of some type of particle (neutrons, alpha radiation, etc.) or energy (photons, heat, etc.) crossing a unit area per unit time. The unit of flux is the number of particles, energy, etc., per square centimeter per second. (2)
food irradiation. Food irradiation is a technology for controlling spoilage and eliminating food-borne pathogens, such as salmonella. The result is similar to conventional pasteurization and is often called "cold pasteurization" or "irradiation pasteurization." Like pasteurization, irradiation kills bacteria and other pathogens that could otherwise result in spoilage or food poisoning. The fundamental difference between the two methods is the source of the energy they rely on to destroy the microbes. Conventional pasteurization relies on heat; irradiation relies on the energy of ionizing radiation. No preservation method, according to the U.S. Food and Drug Administration, is a substitute for safe food handling procedures. (10)
force-on-force (FOF). Security exercise that consists of mock combat between adversary forces and security forces. Adversary forces attempt to reach and damage key safety systems and components of a nuclear reactor. The security force seeks to stop the adversaries from reaching the plant's equipment and causing a release of radiation.
formula quantity. Strategic special nuclear material in any combination in a quantity of 5,000 grams or more computed by the formula, grams = (grams contained U-235) + 2.5 (grams U-233 + grams plutonium). This class of material is sometimes referred to as a category I quantity of material. (2)
fuel assembly. A cluster of fuel rods (or plates). Also called a fuel element. Many fuel assemblies make up a reactor core. (2)
fuel cycle. The series of steps involved in supplying fuel for nuclear power reactors. It can include mining, milling, isotopic enrichment, fabrication of fuel elements, use in a reactor, chemical reprocessing to recover the fissionable material remaining in the spent fuel, reenrichment of the fuel material, refabrication into new fuel elements, and waste disposal. (2)
fuel reprocessing. The processing of reactor fuel to separate the unused fissionable material from waste material. (2)
fuel rod. A long, slender tube that holds fuel for nuclear reactor use. Fuel rods are assembled into bundles called fuel elements or fuel assemblies, which are loaded individually into the reactor core. (2)
fuel temperature coefficient of reactivity. The change in reactivity per degree change in the fuel temperature. The physical property of fuel pellet material (uranium-238) that

causes the uranium to absorb more neutrons away from the fission process as fuel pellet temperature increases. It stabilizes power reactor operations. This coefficient is also known as the Doppler coefficient. (2)

fusion reaction. A reaction in which at least one heavier, more stable nucleus is produced from two lighter, less stable nuclei. Reactions of this type are responsible for enormous release of energy, as in the energy of stars, for example. (2)

gamma radiation. High-energy, short wavelength, electromagnetic radiation emitted from the nucleus. Gamma radiation frequently accompanies alpha and beta emissions and always accompanies fission. Gamma rays are very penetrating and are best stopped or shielded by dense materials, such as lead or depleted uranium. Gamma rays are similar to x-rays. (2)

gap. The space inside a reactor fuel rod that exists between the fuel pellet and the fuel rod cladding. (2)

gas. A substance possessing perfect molecular mobility and the property of indefinite expansion, as opposed to a solid or liquid; any such fluid or mixture of fluids other than air. Normally, these formless substances completely fill the space, and take the shape of, their container. (2)

gas centrifuge. A uranium enrichment process that uses a large number of rotating cylinders in a series to separate uranium-235 (U-235) from uranium-238. Significantly more U-235 enrichment can be obtained from a single unit gas centrifuge than from a single unit gaseous diffusion barrier. Although no gas centrifuge plants are operating in the United States, Louisiana Energy Services (LES) and the U.S. Enrichment Corporation (USEC) had plans to submit license applications in 2002 and 2004, respectively. (2)

gas-cooled reactor. A nuclear reactor in which a gas is the coolant. (2)

gaseous diffusion plant. A facility where uranium hexafluoride gas is filtered. Uranium-235 is separated from uranium-238, increasing the percentage of uranium-235 from 1% to about 3%. The process requires enormous amounts of electric power. (2)

Geiger-Mueller counter. The most commonly used portable radiation detector and measuring instrument. When ionizing radiation passes through the gas-filled tube, a short, intense pulse of current passes from the negative electrode to the positive electrode and is measured or counted. The number of pulses per second measures the intensity of the radiation field.

generation (gross). The total amount of electric energy produced by a generating station as measured at the generator terminals. (2)

generation (net). The gross amount of electric energy produced less the electric energy consumed at a generating station for station use. (2)

gigawatt. One billion watts. (2)

gigawatt hour. One billion watt-hours. (2)

Global Nuclear Energy Partnership (GNEP). Part of President George W. Bush's Advanced Energy Initiative. Its stated purpose is "to develop worldwide consensus on enabling expanded use of economical, carbon-free nuclear energy to meet growing electricity demand. This will use a nuclear fuel cycle that enhances energy security, while promoting non-proliferation. It would achieve its goal by having nations with secure, advanced nuclear capabilities provide fuel services—fresh fuel and recovery of used fuel—to other nations who agree to employ nuclear energy for power generation purposes only. The closed fuel cycle model envisioned by this partnership requires development and deployment of technologies that enable recycling and consumption of long-lived radioactive waste."(11)

global warming. An increase in the near surface temperature of the earth caused by human emissions of greenhouse gases (such as carbon dioxide, methane, ozone, nitrous oxide, and water vapor), which trap the sun's heat in the earth's atmosphere.

GNEP. *See* Global Nuclear Energy Partnership.

graphite. A form of carbon, similar to that used in pencils, used as a moderator in some nuclear reactors. (2)

gray (Gy). The international system (SI) unit of absorbed dose. One gray is equal to an absorbed dose of 1 joule/kilogram (1 Gy = 100 rads). (2)

greater than class C waste or GTCC waste. Low-level radioactive waste that exceeds the concentration limits of radionuclides established for class C waste in 10 CFR 61.55. (2)

green building. Green, or sustainable, building is the practice of creating healthier and more resource-efficient models of construction, renovation, operation, maintenance, and demolition. (12)

greenhouse gas. A gas, such as water vapor, carbon dioxide, nitrous oxide, methane, hydrofluorocarbons (HFCs), perfluorocarbons (PFCs), and sulfur hexafluoride, that is transparent to solar (short-wave) radiation but opaque to long-wave (infrared) radiation, thus preventing long-wave radiant energy from leaving the earth's atmosphere. The net effect is a trapping of absorbed radiation and a tendency to warm the planet's surface. (3)

half-life. The average time for half the initial radioactivity of a radioisotope to have decayed. By the time seven half-lives have elapsed, the radioactivity will have declined to about 1% of the original activity. Measured half-lives vary from millionths of a second to billions of years. Also called physical or radiological half-life. (2)

half-life, biological. The time required for a biological system, such as that of a human, to eliminate, by natural processes, half of the amount of a substance (such as a radioactive material) that has entered it. (2)

half-life, effective. The time required for the amount of a radioactive element deposited in a living organism to decline by one half as a result of the combined action of radioactive decay and biological elimination. (2)

half-thickness. Any given absorber that will reduce the intensity of an original beam of ionizing radiation to one half of its initial value. (2)

hazard. (*a*) Potential for radiation, a chemical, or other pollutant to cause human illness or injury. (*b*) In the pesticide program, the inherent toxicity of a compound. Hazard identification of a given substances is an informed judgment based on verifiable toxicity data from animal models or human studies. (4)

head, reactor vessel. The removable top section of a reactor pressure vessel. It is bolted in place during power operation and removed during refueling to permit access of fuel-handling equipment to the core. (2)

health physics. The science concerned with the evaluation and control of health and environmental hazards that may arise from the use and application of ionizing radiation. (2)

heap leach. A method of extracting uranium from ore using a leaching solution. Small ore pieces are placed in a heap on an impervious material (plastic, clay, asphalt) with perforated pipes under the heap. Acidic solution is then sprayed over the ore, dissolving the uranium. The solution in the pipes is collected and transferred to an ion-exchange system for concentration of the uranium. (2)

heat exchanger. Any device that transfers heat from one fluid (liquid or gas) to another fluid or to the environment. (2)

heat sink. Anything that absorbs heat. It is usually part of the environment, such as the air, a river, or a lake. (2)

heatup. The rise in temperature of the reactor fuel rods resulting from an increase in the rate of fission in the core. (2)

heavy water (D_2O). Water containing a higher than normal proportion (1 in 6,500) of deuterium (heavy hydrogen) atoms. Heavy water is used as a moderator in some nuclear reactors.

heavy water moderated reactor. A reactor that uses heavy water as its moderator and natural uranium as its fuel. These reactors can be used to produce plutonium and tritium. (2)

HEU. *See* highly enriched uranium.

high radiation area. Any area with dose rates greater than 100 millirems (1 millisievert) in 1 hour at a distance 30 centimeters from the ionizing radiation. Access into these areas is maintained under strict control.

high-level waste (HLW). Radioactive materials at the end of a useful life cycle that should be properly disposed of, including (*a*) the highly radioactive material resulting from the reprocessing of spent nuclear fuel, (*b*) irradiated reactor fuel, and (*c*) other highly radioactive material that the U.S. Nuclear Regulatory Commission—consistent with existing law—determines by rule require permanent isolation. High-level waste (HLW) can be in the form of spent fuel discharged from commercial nuclear power reactors and in liquid or solid forms from the production of nuclear weapons. (2)

highly enriched uranium (HEU). Uranium enriched to 20% or greater in the isotope uranium-235. (2)

hot. A colloquial term meaning highly radioactive. (2)

hot spot. The region in a radiation/contamination area where the level of radiation/contamination is significantly greater than in neighboring regions in the area.

hydrological monitoring. Continual monitoring of water at and around nuclear sites so that if any radioactive materials are released from the disposal facility during operations they will be detected. This involves both surface water and groundwater. *See* "Monitoring and Surveillance of Nuclear Waste Sites."

IAEA. *See* International Atomic Energy Agency.

in situ leach. A process using a leaching solution to extract uranium from underground ore bodies in place. The leaching agent, which contains an oxidant such as oxygen with sodium carbonate, is injected through wells into the ore body in a confined aquifer to dissolve the uranium. This solution is then pumped via other wells to the surface for processing. (2)

in vitro. Testing or action outside an organism (e.g., inside a test tube or culture dish). (9)

in vivo. Testing or action inside an organism. (4)

independent spent fuel storage installation (ISFSI). A complex constructed for the interim storage of spent nuclear fuel, solid reactor-related GTCC (greater than class C) waste, and other radioactive waste materials. An ISFSI may be considered an independent facility by the U.S. Nuclear Regulatory Commission even when located on the site of another facility licensed by the commission (such as a nuclear power plant).

indicator. (*a*) In biology, any biological entity or processes or community whose characteristics show the presence of specific environmental conditions. (*b*) In chemistry, a substance that shows a visible change, usually of color, at a desired point in a chemical reaction. (*c*) A device that indicates the result of a measurement; e.g., a pressure gauge or a moveable scale. (4)

individual plant examination (IPE). A risk analysis that considers the unique aspects of a particular nuclear power plant, identifying the specific vulnerabilities to severe accident of that plant. (2)

individual plant examination for external events (IPEEE). While the "individual plant examination" takes into account events that could challenge the design from things that could go awry internally (in the sense that equipment might fail because components do not work as expected), the "individual plant examination for external events" considers challenges such as earthquakes, internal fires, and high winds. (2)

induced radioactivity. Radioactivity that is created when stable substances are bombarded by ionizing radiation. For example, the stable isotope cobalt-59 becomes the radioactive isotope cobalt-60 under neutron bombardment. (2)

Institute of Nuclear Power Operations (INPO). Agency created by the nuclear power industry in 1979 to promote operational excellence in safety and reliability of nuclear power plants. INPO inspects member nuclear power plants regularly and issues recommendations for operational improvement.

institutional control. A "passive" surveillance method used to keep hazardous materials from getting out of a site and to keep people from getting into a site, including deed restrictions and signs and ownership retention. *See* "Monitoring and Surveillance of Nuclear Waste Sites."

integrated gasified combined cycle technology (IGCCT). Coal, water, and oxygen are fed to gasifier, which produces syngas. This medium-Btu gas is cleaned (particulates and sulfur compounds removed) and fed to a gas turbine. The hot exhaust of the gas turbine and heat recovered from the gasification process are routed through a heat-recovery generator to produce steam, which drives a steam turbine to produce electricity. (3)

integrated plant evaluation. An evaluation that considers the plant as a whole rather than system by system. (2)

Intergovernmental Panel on Climate Change (IPCC). Established in 1988 by the World Meteorological Organization and the United Nations Environmental Program to assess the scientific information relating to climate change and to formulate realistic response strategies. (3)

International Atomic Energy Agency (IAEA). The world's center of cooperation in the nuclear field, set up as the world's "Atoms for Peace" organization in 1957 within the United Nations family. The agency works with its member states and multiple partners worldwide to promote safe, secure, and peaceful nuclear technologies. The IAEA Secretariat is headquartered at the Vienna International Centre in Vienna, Austria. Operational liaison and regional offices are located in Geneva, Switzerland; New York City; Toronto; and Tokyo. The IAEA runs or supports research centers and scientific laboratories in Vienna and Seibersdorf, Austria; Monaco; and Trieste, Italy. (13)

International Commission on Radiological Protection (ICRP). Advisory body created in 1928 to provide recommendations on radiation protection. According to its mission statement, the ICRP is an "independent Registered Charity, established to advance for the public benefit the science of radiological protection, in particular by providing recommendations and guidance on all aspects of protection against ionizing radiation."(14)

iodine. Iodine (chemical symbol I) is a nonmetallic solid element. There are both radioactive and nonradioactive isotopes of iodine. Iodine-129 and -131 are the most important radioactive isotopes in the environment. Some isotopes of iodine, such as I-123 and I-124 are used in medical imaging and treatment but are generally not a problem in the environment because they have very short half-lives. (10)

iodine spiking factor. The magnitude of a rapid, short-term increase in the appearance rate of radioiodine in the reactor coolant system. This increase is generally caused by a reactor transient that results in a rapid drop in reactor coolant system pressure relative to the fuel rod internal pressure. (2)

ion. (*a*) An atom that has too many or too few electrons, causing it to have an electrical charge, and therefore, be chemically active. (*b*) An electron that is not associated (in orbit) with a nucleus. (2)

ion-exchange. A common method for concentrating uranium from a solution. The uranium solution is passed through a resin bed where the uranium-carbonate complex ions are transferred to the resin by exchange with a negative ion like chloride. After build-up of the uranium complex on the resin, the uranium is eluted with a salt solution and the uranium is precipitated in another process. (2)

ionization. The process of adding or removing electrons from atoms or molecules, thereby creating ions. High temperatures, electrical discharges, or nuclear radiations can cause ionization. (2)

ionization chamber. An instrument that detects and measures ionizing radiation by measuring the electrical current that flows when radiation ionizes gas in a chamber, making the gas a conductor of electricity. (2)

ionizing radiation. Any radiation capable of displacing electrons from atoms or molecules, thereby producing ions. Some examples are alpha, beta, gamma, x-rays, neutrons, and ultraviolet light. High doses of ionizing radiation may produce severe skin or tissue damage. (2)

irradiation. Exposure to radiation. (2)

isotope. A variant of an element having the same number of protons in its nucleus as other varieties of an element but a different number of neutrons.

isotope separation. The process of separating isotopes from one another, or changing their relative abundances, as by gaseous diffusion or electromagnetic separation. Isotope separation is a step in the isotopic enrichment process. (2)

isotopic enrichment. A process by which the relative abundance of the isotopes of a given element are altered, producing a form of the element that has been enriched in one particular isotope and depleted in its other isotopic forms. (2)

joule. The meter-kilogram-second unit of work or energy, equal to the work done by a force of one newton when its point of application moves through a distance of one meter in the direction of the force; equivalent to 107 ergs and one watt-second. (3)

kilo-. A Greek prefix meaning "thousand" in the nomenclature of the metric system that multiplies a basic unit by 1,000. (2)

kilovolt. The unit of electrical potential equal to 1,000 volts. (2)

kinetic energy. The energy that a body possesses by virtue of its mass and velocity. Also called the energy of motion. (2)

laser enrichment technology. A technology used to enrich natural UF6 gas in the uranium-235 isotope. *See* "Sustainability."

lens dose equivalent. The external exposure dose equivalent to the lens of the eye at a tissue depth of 0.3 centimeters (300 mg/cm^2). (2)

lethal dose (LD). The dose of radiation expected to cause death to 50% of an exposed population within 30 days (LD 50/30). Typically, the LD 50/30 is in a range of 400 to 450 rem (4 to 5 sieverts) received over a very short period. (2)

license termination plan (LTP). For a nuclear power plant, describes the site in detail and includes plans for dismantling the facilities, plans for site remediation, plans for a radiation survey, plans for the end use of the site, and updates on estimated decommissioning costs and environmental impact.

licensed material. Source material, special nuclear material, or byproduct material received, possessed, used, transferred, or disposed of under a general or specific license issued by the U.S. Nuclear Regulatory Commission. (2)

licensee (as in U.S. Nuclear Regulatory Commission licensee). A public or private utility company that has received permits from the U.S. Nuclear Regulatory Commission to construct or operate a nuclear power plant. (9)

licensing basis. The collection of documents or technical criteria that provides the basis on which the U.S. Nuclear Regulatory Commission issues a license to possess radioactive materials, conduct operations involving emission of radiation, use special nuclear materials, or dispose of radioactive waste. (2)

life-cycle cost analysis (LCCA). A process for evaluating the total economic worth of a project by analyzing initial costs and discounted future costs, such as maintenance, user, reconstruction, rehabilitation, restoring, and resurfacing costs, over the life of the project.

light water. Ordinary water as distinguished from heavy water. (2)

light water reactor. A term used to describe reactors using ordinary water as coolant, including boiling water reactors (BWRs) and pressurized water reactors (PWRs), the most common types used in the United States. (2)

limiting condition for operation. The section of technical specifications that identifies the lowest functional capability or performance level of equipment required for safe operation of the facility. (2)

limiting safety system settings. Settings for automatic protective devices related to those variables having significant safety functions. Where a limiting safety system setting is specified for a variable on which a safety limit has been placed, the setting will ensure that automatic protective action will correct the abnormal situation before a safety limit is exceeded. (2)

linear energy transfer (LET). Measurement of the energy transferred to a medium from ionizing radiation as the radiation travels through a medium, per unit distance traveled. Gamma rays have low LET; alpha rays have high LET.

linear heat generation rate. The heat generation rate per unit length of fuel rod, commonly expressed in kilowatts per foot (kw/ft) of fuel rod. (2)

linear no threshold hypothesis (LNTH). An extrapolation from well-established health effects of radiation at both low and high doses, used for setting radiation-protection standards worldwide.

long-term surveillance and maintenance (LTSM). Activities carried out at closed nuclear sites to protect human and environmental health, consisting of engineering and institutional controls.

loop. In a pressurized water reactor (PWR), the coolant flow path through piping from the reactor pressure vessel to the steam generator to the reactor coolant pump and back to the reactor pressure vessel. Large PWRs may have as many as four separate loops. (2)

loss of coolant accident (LOCA). Those postulated accidents that result in a loss of reactor coolant at a rate in excess of the capability of the reactor makeup system from breaks in the reactor coolant pressure boundary, up to and including a break equivalent in size to the double-ended rupture of the largest pipe of the reactor coolant system. (2)

low population zone (LPZ). An area of low population density often required around a nuclear installation before it is built. The number and density of residents is of concern in emergency planning so that certain protective measures (such as notification and instructions to residents) can be accomplished in a timely manner. (2)

low specific activity (LSA). Radioactive material with limited specific activity that is nonfissile or is excepted under 10 CFR 71.15. Shielding materials surrounding the LSA material may not be considered in determining the estimated average specific activity of the package contents. (15)

lower explosive limit (LEL). The minimum concentration of gas or vapor in air below which a substance does not burn when exposed to an ignition source. (6)

low-level waste. A general term for a wide range of wastes having low levels of radioactivity. Industries, medical and research facilities, nuclear power reactors, and nuclear fuel fabrication plants generate low-level wastes as part of their normal operations. These wastes are generated in many physical and chemical forms and levels of contamination. Low-level radioactive wastes containing source, special nuclear, or byproduct material are acceptable for disposal in a land disposal facility. For the purposes of this definition, low-level waste has the same meaning as in the Low-Level Radioactive Waste Policy Act, that is, radioactive waste not classified as high-level radioactive waste, transuranic waste, spent nuclear fuel, or byproduct material as defined in section 11e. (2)

Manhattan Project. The U.S. government project that produced the first nuclear weapons during World War II. Started in 1942, the Manhattan Project formally ended in 1946. The Hanford Site (WA), Oak Ridge Reservation (TN), and Los Alamos National Laboratory (NM) were created for this effort. The project was named for the Manhattan Engineer District of the U.S. Army Corps of Engineers. (3)

mass number. The number of neutrons and protons in the nucleus of an atom. Also known as the atomic weight. (2)

mass-energy equation. The equation developed by Albert Einstein, which is usually given as $E = mc^2$, showing that when the energy of a body changes by an amount E (no matter what form the energy takes), the mass (m) of the body will change by an amount equal to E/c^2. The factor c squared, the speed of light in a vacuum (3×10^8), may be regarded as the conversion factor relating units of mass and energy. The equation predicted the possibility of releasing enormous amounts of energy by the conversion of mass to energy. It is also called the Einstein equation. (2)

maximum dependable capacity (gross). In a nuclear power reactor, dependable main-unit gross generating capacity, winter or summer, whichever is smaller. The dependable capacity varies because the unit efficiency varies during the year because of temperature variations in cooling water. It is the gross electrical output as measured at the output terminals of the turbine generator during the most restrictive seasonal conditions (usually summer). (2)

maximum dependable capacity (net). In a nuclear power reactor, gross maximum dependable generating capacity less the normal station service loads. (2)

mega-. A prefix that multiplies a basic unit by 1,000,000 (1×10^6). (2)

megacurie. One million curies. (2)

megawatt (MW). One million watts. (2)

megawatt hour (MWh). One million watt-hours. (2)

metric ton. Approximately 2,200 pounds in the English system of measurements. (In the international system of measurements, 1 metric ton = 1,000 kg.)(2)

micro-. A prefix that divides a basic unit into 1 million parts (0.000001). (2)

microcurie. One millionth of a curie. That amount of radioactive material that disintegrates (decays) at the rate of 37 thousand atoms per second. (2)

mill tailings. Leftover crushed rock from the processing of uranium ore into yellowcake in a mill. Tailings contain several naturally occurring radioactive elements, including uranium, thorium, radium, polonium, and radon, and must be treated as radioactive waste.

milli-. A prefix that divides a basic unit by 1,000. (2)

millirem. One thousandth of a rem (0.001 rem). (2)

milliroentgen (mR). One thousandth of a roentgen (R). 1mR = 10^{-3} R = 0.001 R. (2)

MIRV. *See* multiple independent targeted reentry vehicle.

mixed oxide (MOX) fuel. A mixture of uranium oxide and plutonium oxide used to fuel a reactor. Conventional nuclear fuel is made of pure uranium oxide. (2)

moderator. A material, such as ordinary water, heavy water, or graphite, that is used in a reactor to slow down high-velocity neutrons, thus increasing the likelihood of fission. (2)

moderator temperature coefficient of reactivity. As the moderator (water) increases in temperature, it becomes less dense and slows down fewer neutrons, resulting in a negative change of reactivity. This negative temperature coefficient stabilizes atomic power reactor operations. (2)

molecule. A group of atoms held together by chemical forces. A molecule is the smallest unit of a compound that can exist by itself and retain all of its chemical properties. (2)

Monitored Retrievable Storage (MRS) Installation. A complex designed, constructed, and operated by the U.S. Department of Energy for the receipt, transfer, handling, packaging, possession, safeguarding, and storage of spent nuclear fuel aged for at least 1 year, solidified high-level radioactive waste resulting from civilian nuclear activities, and solid reactor-related GTCC (greater than class C) waste, pending shipment to a high-level waste repository or other disposal. (2)

monitoring of radiation. Periodic or continuous determination of the amount of ionizing radiation or radioactive contamination present in a region, as a safety measure, for the purpose of health or environmental protection. Monitoring is done for air, surface water and groundwater, soil and sediment, equipment surfaces, and personnel (for example, bioassay or alpha scans). (2)

multiple independent targeted reentry vehicle (MIRV). A reentry vehicle carried by a delivery system that can place one or more reentry vehicles over each of several separate targets. (16)

nano-. A prefix that divides a basic unit by one billion (10^{-9}). (2)

nanocurie. One billionth 10^{-9} of a curie. (2)

National Atmospheric Release Advisory Center (NARAC). Located at the University of California' s Lawrence Livermore National Laboratory, NARAC provides services to the U.S. government that map the probable spread of hazardous material, including nuclear, chemical, or biological, accidentally or intentionally released into the atmosphere. NARAC provides predictions in real time during an emergency that can inform a decision to take protective action for the health and safety of people in affected areas.

natural circulation. The circulation of coolant in the reactor coolant system, without the use of pumps, due to natural convection resulting from the different densities of relative cold and heated portions of the system.

natural uranium. Uranium as found in nature. It contains 0.7% uranium-235, 99.3% uranium-238, and a trace of uranium-234 by weight. In terms of the amount of radioactivity, it contains approximately 2.2% uranium-235, 48.6% uranium-238, and 49.2% uranium-234. (2)

net summer capability. The steady hourly output that generating equipment is expected to supply to system load exclusive of auxiliary power, as demonstrated by tests at the time of summer peak demand. (2)

neutron. An uncharged elementary particle, with a mass slightly greater than that of the proton, found in the nucleus of every atom heavier than hydrogen. (2)

neutron, thermal. A neutron that has (by collision with other particles) reached an energy state equal to that of its surroundings, typically on the order of 0.025 eV (electron volts). (2)

neutron capture. The reaction that occurs when a nucleus captures a neutron. The probability that a given material will capture a neutron is proportional to its neutron capture cross-section and depends on the energy of the neutrons and the nature of the material. (2)

neutron chain reaction. A process in which some of the neutrons released in one fission event cause other fissions to occur. There are three types of chain reactions: (*a*) nonsustaining, in which the number of fissions will decline to zero, (*b*) sustaining, in which the number of fissions will remain constant, and (*c*) multiplying, in which the number of fissions will increase. (2)

neutron flux. A measure of the intensity of neutron radiation in neutrons/cm^2 per second. It is the number of neutrons passing through 1 square centimeter of a given target in 1 second. Expressed as nv, where n = the number of neutrons per cubic centimeter and v = their velocity in centimeters per second.

neutron generation. The release and absorption of fission neutrons by a fissile material and the fission product of that material producing a second generation of neutrons. In a typical nuclear power reactor system, there are about 40,000 generations of neutrons every second. (2)

neutron leakage. Neutrons that escape from the vicinity of the fissionable material in a reactor core. Neutrons that leak out of the fuel region are no longer available to cause fission and must be absorbed by shielding placed around the reactor pressure vessel for that purpose. (2)

neutron poison. In reactor physics, a material other than fissionable material in the vicinity of the reactor core that will absorb neutrons. The addition of poisons, such as control rods or boron, into the reactor is said to be an addition of negative reactivity. (2)

neutron source. Any material that emits neutrons, such as a mixture of radium and beryllium, that can be inserted into a reactor to ensure a neutron flux large enough to be distinguished from background to register on neutron detection equipment. (2)

noble gas. A gaseous element that does not readily combine chemically with other elements. An inert gas. Examples are helium, argon, krypton, xenon, and radon. (2)

nonpower reactor. Reactors used for research, training, and test purposes and for the production of radioisotopes for medical and industrial uses. (2)

nonproliferation. Preventing or limiting the spread of nuclear weapons. A Nuclear Non Proliferation Treaty has been signed by 189 countries.

nonstochastic effect. Inevitable or highly probable health effects of radiation, the severity of which varies with doses over a threshold amount. Radiation-induced cataract formation is an example of a nonstochastic effect (also called a deterministic effect). (2)

nonvital plant systems. Systems at a nuclear facility that may or may not be necessary for the operation of the facility (i.e., power production) but that would have little or no effect on public health and safety should they fail. These systems are not safety related. (2)

not applicable (NA). Specifies that a particular field is not applicable to the event. (2)

not reported (NR). Specifies that information applicable to the particular field was not included in the event report. (2)

nozzle. As used in power water reactors and boiling water reactors, the interface (inlet and outlet) between reactor plant components (pressure vessel, coolant pumps, steam generators, etc.) and their associated piping systems. (2)

nuclear energy. The energy liberated by a nuclear reaction (fission or fusion) or by radioactive decay. (2)

Nuclear Energy Agency (NEA). A specialized agency within the Organization for Economic Co-operation and Development (OECD), an intergovernmental organization of industrialized countries, based in Paris, France. (17)

nuclear force. A powerful short-ranged attractive force that holds together the particles inside an atomic nucleus. (2)

Nuclear Material Management and Safeguards System (NMMSS). Standardized database run by the U.S. Department of Energy to track the possession, use, and shipment of nuclear materials at government and commercial nuclear facilities.

Nuclear Nonproliferation Treaty (NPT). A multilateral treaty signed in 1968 whose goal is to limit the spread of nuclear weapons, promote cooperation in the use of nuclear technology, and further the goal of achieving nuclear disarmament; 190 states have joined the treaty, including the 5 states that acknowledge their possession of nuclear weapons.

nuclear power plant. An electrical generating facility using a nuclear reactor as its heat source to provide steam to a turbine generator. (2)

Nuclear Regulatory Commission (NRC). An independent federal regulatory agency, created by Congress in 1974, responsible for regulating civilian use of nuclear materials; licensing and inspecting nuclear power plants, universities and other facilities using radioactive materials; and regulating the disposal of nuclear waste.

Nuclear Regulatory Commission (NRC) Operations Center. The focal coordination point in Rockville, Maryland, for communicating with NRC licensees, state agencies, and other federal agencies about operating events in both the nuclear reactor and nuclear material industry. The center is staffed 24 hours a day by an NRC headquarters operations officer (HOO), who is trained to receive, evaluate, and respond to events reported to the center. (2)

nuclear steam supply system. The reactor and the reactor coolant pumps (and steam generators for a pressurized water reactor) and associated piping in a nuclear power plant used to generate the steam needed to drive the turbine generator unit. (2)

nuclear waste. A particular type of radioactive waste that is produced as part of the nuclear fuel cycle. These include extraction of uranium from ore, concentration of uranium, processing into nuclear fuel, and disposal of byproducts. Radioactive waste is a broader term that includes all waste that contains radioactivity. Residues from water treatment, contaminated equipment from oil drilling, and tailings from the processing of metals such as vanadium and copper also contain radioactivity but are not "nuclear waste" because they are produced outside of the nuclear fuel cycle. The U.S. Nuclear Regulatory Commission generally regulates only those wastes produced in the nuclear fuel cycle (uranium mill tailings, depleted uranium, spent fuel rods, etc.). (2)

nucleon. Common name for particles included in the atomic nucleus. At present, applied to protons and neutrons but may include any other particles found to exist in the nucleus. (2)

nucleus. The small, central, positively charged region of an atom. Except for the nucleus of ordinary hydrogen, which has only a proton, all atomic nuclei contain both protons and neutrons. The number of protons determines the total positive charge or atomic number. This number is the same for all the atomic nuclei of a given chemical element. (2)

nuclide. A general term referring to all known isotopes, both stable (279) and unstable (about 2,700), of the chemical elements. (2)

Oak Ridge Institute. A U.S. Department of Energy institute in Tennessee that focuses on scientific initiatives to research health risks from occupational hazards, assess environmental cleanup, respond to radiation medical emergencies, support national security and emergency preparedness, and educate the next generation of scientists. (18)

occupational dose. The dose received by an individual in the course of employment in which the individual's assigned duties involve exposure to radiation or to radioactive material from licensed and unlicensed sources of radiation, whether in the possession of the licensee or other person. (2)

open pit mining. One of several methods used for mining uranium whereby the overburden is removed and equipment is used to shovel rock containing the ore. *See* "Sustainability."

operable. Describes a system, subsystem, train, component, or device when it is capable of performing its specified functions and when all necessary attendant instrumentation, controls, electrical power, cooling or seal water, lubrication, or other auxiliary equipment that are required for the system, subsystem, train, component, or device to perform its functions are also capable of performing their related support functions. (2)

operational mode. The mode in a nuclear power reactor that corresponds to any one inclusive combination of core reactivity condition, power level, and average reactor coolant temperature. (2)

operational monitoring. A method for monitoring nuclear waste sites that is designed to assess basic operations of a site, such as receipt inspection, offloading from transportation equipment, and disposal of waste. *See* "Monitoring and Surveillance of Nuclear Waste Sites."

orphan sources. *See* Unwanted radioactive material.

packaging. Packaging means the assembly of components necessary to retain the radioactive material during transport. It may consist of one or more receptacles, absorbent materials, spacing structures, thermal insulation, radiation shielding, and devices for cooling or absorbing mechanical shocks. The vehicle, tie-down system, and auxiliary equipment may be designated as part of the packaging. (2)

parent. A radionuclide that upon radioactive decay or disintegration yields a specific nuclide (the daughter). (2)

particle accelerator. Equipment used to split the nucleus of an atom, used in the manufacturing of radionuclides for use in medical procedures. *See* "Radionuclides."

parts per million (ppm). Parts (molecules) of a substance contained in a million parts of another substance (e.g., water). (2)

pellet, fuel. As used in pressurized water reactors and boiling water reactors, a small cylinder approximately 3/8-inch in diameter and 5/8-inch in length, consisting of uranium fuel in a ceramic form—uranium dioxide, UO_2. Typical fuel pellet enrichments in nuclear power reactors range from 2.0% to 3.5% uranium-235. (2)

performance-based regulation. Regulations that state required results or outcomes of performance rather than designating process, technique, or procedure.

performance-based regulatory action. Licensee attainment of defined objectives and results without detailed direction from the U.S. Nuclear Regulatory Commission on how these results are to be obtained. (2)

periodic table. An arrangement of chemical elements in order of increasing atomic number. Elements of similar properties are placed one under the other, yielding groups or families of elements. (2)

personnel monitoring. The use of portable survey meters to determine the amount of radioactive contamination on individuals, or the use of dosimetry to determine an individual's occupational radiation dose. (2)

photon. A quantum (or packet) of energy emitted in the form of electromagnetic radiation. Gamma rays and x-rays are examples of photons. (2)

pico-. A prefix that divides a basic unit by one trillion (10^{-12}). (2)

picocurie. One trillionth (10^{-12}) of a curie. (2)

pig. A colloquial term describing a container (usually lead or depleted uranium) used to ship or store radioactive materials. The thick walls of this shielding device protect the person handling the container from radiation. Large containers used for spent fuel storage are commonly called casks. (2)

pile. A colloquial term describing the first nuclear reactors. They are called piles because the earliest reactors were "piles" of graphite and uranium blocks. (2)

pipe overpack. Used for shipment of waste containing higher concentrations of plutonium, this special container is used to separate materials whose nucleus has the potential of splitting into two or more nuclei, further releasing energy. The overpack is first placed within an impact limiter before it is enclosed in a 55-gallon drum and placed in a containment vessel. *See* "Transportation of Nuclear Waste."

planned special exposure. An infrequent exposure to radiation, separate from and in addition to the annual dose limits (see 10 CFR 20.1003 and 20.1206). (2)

plausible accidents. Postulated events that meet a probability test rather than the more challenging test represented by a design-basis event. (2)

plume exposure pathway EPZ. One of two emergency planning zones or EPZs around a nuclear power plant, the plume exposure pathway EPZ has a radius of about 10 miles from the reactor site. Predetermined protective action plans are in place for this EPZ and are designed to avoid or reduce dose from potential exposure of radioactive materials. These actions include sheltering, evacuation, and the use of potassium iodide where appropriate. (8)

plutonium (Pu). A heavy, radioactive, manmade metallic element with atomic number 94. Its most important isotope is fissile plutonium-239, which is produced by neutron irradiation of uranium-238. It exists in only trace amounts in nature. (2)

pocket dosimeter. A small ionization detection instrument that indicates ionizing radiation exposure directly. (2)

pool reactor. A reactor in which the fuel elements are suspended in a pool of water that serves as the reflector, moderator, and coolant. Popularly called a "swimming pool reactor," it is used for research and training, not for electrical generation. (2)

positron. Particle equal in mass to the electron but opposite in charge. A positive electron. (2)

positron emission tomography (PET). A procedure in which a small amount of radioactive glucose (sugar) is injected into a vein, and a scanner is used to make detailed, computerized pictures of areas inside the body where the glucose is used. Because cancer cells often use more glucose than normal cells, the pictures can be used to find cancer cells in the body. Also called PET scan. (19)

possession-only license. A form of license that allows a licensee to possess but not operate a nuclear facility. (2)

post-closure phase. Designated period beginning with the end of the decommissioning phase ("closure") and extending through the end of the regulatory time frame established for the specific nuclear material repository. The maximum allowable radiation exposure to humans within a radius of the repository during this period is established by the U.S. Environmental Protection Agency (EPA). For the proposed geological repository at Yucca Mountain (NV), EPA has established maximum limits for a human within 10 miles of the facility for the first 10,000 years and from 10,000 to 1,000,000 years. (20)

post-shutdown decommissioning activities report (PSDAR). Report to be submitted within 2 years of ending nuclear power plant site operations that must include a description of the planned decommissioning activities, a schedule for these activities, an estimate of expected cost, and a review of environmental impacts associated with decommissioning.

potassium iodide. Potassium iodide (KI) is a salt, similar to table salt. It is routinely added to table salt to make it "iodized." Ingestion of potassium iodide can block the thyroid gland's uptake of radioactive iodine, reducing the risk from radiation exposure of thyroid cancers and other diseases that might otherwise be caused by exposure to radioactive iodine that could be dispersed in a severe nuclear accident. (21)

power coefficient of reactivity. The change in reactivity per percent change in power. The power coefficient is the summation of the moderator temperature coefficient of reactivity, the fuel temperature coefficient of reactivity, and the void coefficient of reactivity. (2)

power defect. The total amount of reactivity added because of a given change in power. It can also be expressed as the integrated power coefficient over the range of the power change. (2)

power reactor. A reactor designed to produce heat for electric generation (as distinguished from reactors used for research), for producing radiation or fissionable materials or for reactor component testing. (2)

preliminary notification (PN). A brief summary report issued by the U.S. Nuclear Regulatory Commission staff to notify the commission of the occurrence of a significant event that appears to have health and safety significance or major public or media interest. PNs are based on information provided by state radiation control program staff. (2)

pressure vessel. A strong-walled container housing the core of most types of power reactors. It usually also contains the moderator, neutron reflector, thermal shield, and control rods. (2)

pressurized water reactor (PWR). A power reactor in which heat is transferred from the core to an exchanger by high temperature water kept under high pressure in the primary system. Steam is generated in a secondary circuit. Many reactors producing electric power are pressurized water reactors. (2)

pressurizer. A tank or vessel that controls the pressure in a pressurized water reactor. (2)

primary system. A term that may be used to refer to the reactor coolant system. (2)

primordial radionuclides. A category of naturally occurring radionuclides that have extraordinarily long half-lives (uranium-238, thorium-232, and potassium 40). *See* "Radionuclides."

probabilistic risk analysis (PRA). A systematic and comprehensive methodology to evaluate risks associated with a complex engineered technological entity. Consequences are expressed numerically (e.g., the number of people potentially hurt or killed) and their likelihoods of occurrence are expressed as probabilities or frequencies (i.e., the number of occurrences or the probability of occurrence per unit time). The total risk is the sum of the products of the consequences multiplied by their probabilities. PRA is used to address the risk triplet, the set of three questions that the U.S. Nuclear Regulatory Commission (NRC) uses to define risk: What can go wrong? How likely is it? and What are the consequences? The NRC identifies important scenarios from such an assessment. (2)

production expense. Production expenses are a component of generation expenses that includes costs associated with operation, maintenance, and fuel. (2)

Project Management Information System (PMIS). A tool the Office of Information Services at the Nuclear Regulatory Commission is implementing to support internal stakeholders, PMIS collects all relevant information pertaining to an application or system development project and once fully implemented, the system will be the primary communication tool used to inform internal stakeholders of progress and issues related to information technology projects. (22)

proportional counter. A radiation instrument in which an electronic detection system receives pulses that are proportional to the number of ions formed in a gas-filled tube by ionizing radiation. (2)

proprietary information. Privately owned knowledge or data, such as that protected by a registered patent, copyright, or trademark. (2)

proton. An elementary nuclear particle with a positive electric charge located in the nucleus of an atom. (2)

Prussian blue. A substance that helps speed the body's removal of radioactive cesium-137 by trapping it within the intestines for excretion and reducing the half-life from 110 to 30 days.

public dose. The dose received by a member of the public from exposure to radiation or to radioactive material released by a licensee, or to any other source of radiation under the control of a licensee. Public dose does not include occupational doses, background radiation, or radiation from individual medical procedures. (2)

quality assurance/quality control (QA/QC). A system of procedures, checks, audits, and corrective actions to ensure that all U.S. Environmental Protection Agency research design and performance, environmental monitoring and sampling, and other technical and reporting activities are of the highest achievable quality. (4)

quality factor. The factor by which the absorbed dose (rad or gray) is to be multiplied to obtain a quantity that expresses the biological damage (rem or sievert) to an exposed individual. It is used because some types of radiation, such as alpha particles, are more biologically damaging internally than other types. (2)

quantitative risk assessment (QRA). As with probabilistic risk analysis, quantitative risk assessment requires the calculations of two components of risk, the magnitude of the potential loss, and the probability that the loss will occur. *See* Probabilistic risk analysis (PRA).

quantum theory. The concept that energy is radiated intermittently in packets of definite magnitude, called quanta, and absorbed in a like manner. (2)

rad. The special unit for radiation absorbed dose, which is the amount of energy from any type of ionizing radiation (e.g., alpha, beta, gamma, neutrons) deposited in any medium (e.g., water, tissue, air). A dose of one rad means the absorption of 100 ergs (a small but measurable amount of energy) per gram of absorbing tissue (100 rad = 1 gray). (2)

radiation, ionizing. Alpha particles, beta particles, gamma rays, x-rays, neutrons, high-speed electrons, high-speed protons, and other particles capable of producing ions. The term radiation generally does not refer to non-ionizing radiation, such as radio- or microwaves, or visible, infrared, or ultraviolet light. (2)

radiation, nuclear. Particles (alpha, beta, neutrons) or photons (gamma) emitted from radioactive nuclei as a result of radioactive decay. (2)

radiation area. Any area with radiation levels greater than 5 millirems (0.05 millisievert) in 1 hour at 30 centimeters from the source or from any surface through which the radiation penetrates. (2)

radiation detection instrument. A device that detects and displays the characteristics of ionizing radiation. (2)

radiation shielding. Reduction of radiation by interposing a shield of absorbing material between any radioactive source and a person, work area, or radiation-sensitive device. (2)

radiation sickness (syndrome). The complex of symptoms characterizing the disease known as radiation injury, resulting from excessive exposure to ionizing radiation. The earliest of these symptoms are nausea, fatigue, vomiting, and diarrhea, which may be followed by loss of hair, hemorrhage, inflammation of the mouth and throat, and general loss of energy. In severe cases death may occur within 2–4 weeks. Those who survive 6 weeks after the receipt of a single large dose of radiation to the whole body may generally be expected to recover. (2)

radiation source. Usually a sealed source of radiation used in teletherapy and industrial radiography as a power source for batteries (as in use in space craft) or in various types of industrial gauges. (2)

radiation standards. Exposure standards, permissible concentrations, rules for safe handling, regulations for transportation, regulations for industrial control of radiation, and control of radioactive material by legislative means. (2)

radiation warning symbol. An officially prescribed symbol (a magenta or black trefoil) on a yellow background that must be displayed where certain quantities of radioactive materials are present or where certain doses of radiation could be received. (2)

radioactive contamination. Deposits of radioactive material in any place where it may harm persons or equipment. (2)

radioactive decay. Decrease in the amount of any radioactive material with the passage of time due to the spontaneous emission of radiation from an atomic nucleus.

Radioactive Material Package Certificate of Compliance (CoC). Certificate issued by the U.S. Nuclear Regulatory Commission if packaging of radioactive materials by a company meets all requirements. *See* "Transportation of Nuclear Waste."

radioactive series. A succession of isotopes, each of which transforms by radioactive disintegration into the next until a stable isotope results. The first member is called the parent, the intermediate members are called daughters, and the final stable member is called the end product. (2)

Radioactive Waste Management Committee (RWMC). An international committee located within the Nuclear Energy Agency of the Organization for Economic Cooperation and Development (OECD) and made up of senior representatives from regulatory authorities, radioactive waste management agencies, policy-making bodies, and research and development institutions. Its purpose is to foster international co-operation in the management of radioactive waste and radioactive materials among the OECD member countries. (23)

radioactivity. The spontaneous emission of radiation, generally alpha or beta particles, often accompanied by gamma rays, from the nucleus of an unstable isotope. Also, the rate at which radioactive material emits radiation. Measured in units of becquerels or disintegrations per second. (2)

radiography. The making of a shadow image on photographic film by the action of ionizing radiation. (2)

radioisotope. An unstable isotope of an element that decays spontaneously, emitting radiation. (2)

radiological dispersal device (RDD). Also known as "dirty bombs," these devices combine radioactive material with conventional explosives to disperse radioactive material over a large area.

radiological exposure device (RED). A device whose purpose is to expose people to radiation, rather than to disperse radioactive material into the air, as would an RDD. An RED could be constructed from unshielded or partially shielded radioactive materials in any form placed in any type of container.

radiological sabotage. Any deliberate act directed against a plant or transport in which an activity licensed pursuant to 10 CFR pt. 73 of the U.S. Nuclear Regulatory Commission's regulations is conducted or against a component of such a plant or transport that could directly or indirectly endanger the public health and safety by exposure to radiation. (2)

radiological survey. The evaluation of the radiation hazards accompanying the production, use, or existence of radioactive materials under a specific set of conditions. Such evaluation customarily includes a physical survey of the disposition of materials and equipment, measurements or estimates of the levels of radiation that may be involved, and a sufficient knowledge of processes affecting these materials to predict hazards resulting from expected or possible changes in materials or equipment. (2)

radiology. That branch of medicine dealing with the diagnostic and therapeutic applications of radiant energy, including x-rays and radioisotopes. (2)

radionuclide. A radioisotope. (2)

radiosensitivity. The relative susceptibility of cells, tissues, organs, organisms, or other substances to the injurious action of radiation. (2)

radiotoxic. Characteristic of a radioactive substance that is toxic to living cells.

radium (Ra). A radioactive metallic element with atomic number 88. As found in nature, the most common isotope has a mass number of 226. It occurs in minute quantities associated with uranium in pitchblende, camotite, and other minerals. (2)

radon (Rn). A radioactive element that is one of the heaviest gases known. Its atomic number is 86. It is a daughter of radium. (2)

radura. A symbol developed to label foods that been treated by irradiation. The label is required by the United States and other countries and should be accompanied by the words "irradiated" or "treated by irradiation."

reaction. Any process involving a chemical or nuclear change. (2)

reactivity. A term expressing the departure of a reactor system from criticality. A positive reactivity addition indicates a move toward supercriticality (power increase). A negative reactivity addition indicates a move toward subcriticality (power decrease). (2)

reactor, nuclear. A device in which nuclear fission may be sustained and controlled in a self-supporting nuclear reaction. The varieties are many, but all incorporate certain features, including fissionable material or fuel, a moderating material (unless the reactor is operated on fast neutrons), a reflector to retain escaping neutrons, provisions for removal of heat, measuring and controlling instruments, and protective devices. The reactor is the heart of a nuclear power plant. (2)

reactor bolshoy mushchosty kanaly (RBMK). The type of pressurized water reactor used at Chernobyl, which uses water as its coolant and graphite as a moderator. *See* "Three Mile Island and Chernobyl."

reactor coolant system. The system used to remove energy from the reactor core and transfer that energy either directly or indirectly to the steam turbine. (2)

reactor trip. A term used by pressurized water reactors for a reactor scram.

records-only site. A former nuclear site that has been officially closed, that has been cleaned up to unrestricted use, and that contains primarily records. *See* "Long-Term Surveillance and Maintenance."

reference man. A hypothetical person with anatomical and physiological characteristics of an average individual that is used in calculations assessing internal dose (also may be called "standard man"). (2)

reflector. A layer of material immediately surrounding a reactor core that scatters back (or reflects) into the core many neutrons that would otherwise escape. The returned neutrons can then cause more fissions and improve the efficiency of the reactor. Common reflector materials are graphite, beryllium, water, and natural uranium. (2)

relative biologic effectiveness (RBE). The RBE of some test radiation (r) compared with x-rays is defined by the ratio D_{250}/D_r where D_{250} and D_r are, respectively, the doses of x-rays and the test radiation required for equal biologic effect. (1).

relative risk assessment. Estimating the risks associated with different stressors or management actions. (4)

rem (roentgen equivalent man). A standard unit that measures the effects of ionizing radiation on humans. The dose equivalent in rems is equal to the absorbed dose in rads multiplied by the quality factor of the type of radiation. (2)

remediation. (*a*) Cleanup or other methods used to remove or contain a toxic spill or hazardous materials from a Superfund site. (*b*) For the Asbestos Hazard Emergency Response program, abatement methods including evaluation, repair, enclosure, encapsulation, or removal of greater than 3 linear feet or square feet of asbestos-containing materials from a building. (4)

remote-handled (RH) waste. Transuranic wastes that have a measured radiation dose rate at the container surface of between 200 mrems per hour and 1,000 rems per hour and, therefore, must be shielded for safe handling. (6)

reracking. Increasing the storage capacity of a spent nuclear (cooling) pool by placing the spent nuclear fuel in a more compact configuration.

restricted area. Any area to which access is controlled for the protection of individuals from exposure to radiation and radioactive materials. (2)

risk. The combined answers to What can go wrong? How likely is it? and What are the consequences?(1)

risk communication. The exchange of information about health or environmental risks among risk assessors and managers, the general public, news media, interest groups, etc. (4)

risk-based decision-making. An approach to regulatory decision-making in which such decisions are based solely on the results of a probabilistic risk analysis. (2)

risk-informed decision-making. An approach to decision making in which insights from probabilistic risk analyses are considered with other engineering insights. (2)

risk-informed regulation. Incorporating an assessment of safety significance or relative risk in the U.S. Nuclear Regulatory Commission regulatory actions. Making sure that the regulatory burden imposed by individual regulations or processes is commensurate with the importance of that regulation or process to protecting public health and safety and the environment. (2)

risk-significant. When used to qualify an object, such as a system, structure, component, accident sequence, or cut set, this term identifies that object as exceeding a predetermined criterion related to its contribution to the risk from the facility being addressed. One that is associated with a level of risk that exceeds a predetermined significance criterion. (2)

roentgen (R). A unit of exposure to ionizing radiation. (2)

rubblization. A decommissioning technique involving demolition and burial of formerly operating nuclear facilities. All equipment from buildings is removed and the surfaces are decontaminated. Above-grade structures are demolished into rubble and buried in the structure's foundation below ground. The site surface is then covered, regraded, and landscaped for unrestricted use. (2)

safe shutdown earthquake. The maximum earthquake potential for which certain structures, systems, and components, important to safety, are designed to sustain and remain functional. (2)

safeguards. In the regulation of domestic nuclear facilities and materials, the use of inventory and accounting programs to verify that all special nuclear material is properly controlled and accounted for, and the physical protection (also referred to as physical security) of equipment and security forces. In the regulatory arena, this term applies to systems and procedures at a facility that are relied on to remain functional during and following design-basis events. Examples of safety-related functions include shutting down a nuclear reactor and maintaining it in a safe shutdown condition. As used by the International

Atomic Energy Agency, safeguards verify that the "peaceful use" commitments made in binding nonproliferation agreements are honored.

safety injection. The rapid insertion of a neutron poison (such as boric acid) into the reactor coolant system to ensure reactor shutdown. (2)

safety limit. A restriction or range placed on important process variables that are necessary to reasonably protect the integrity of the physical barriers that guard against the uncontrolled release of radioactivity. (2)

safety related. In the regulatory arena, this term applies to systems, structures, components, procedures, and controls of a facility or process that are relied on to remain functional during and following design-basis events. Their functionality ensures that key regulatory criteria, such as levels of radioactivity released, are met. Examples of safety related functions include shutting down a nuclear reactor and maintaining it in a safe shutdown condition. (2)

safety-significant. When used to qualify an object, such as a system, component, or accident sequence, this term identifies that object as having an impact on safety that exceeds a predetermined significance criterion. (2)

SAFSTOR (safe storage). A method of decommissioning in which the nuclear facility is placed and maintained in such condition that the nuclear facility can be safely stored and subsequently decontaminated to levels that permit release for unrestricted use. (2)

sarcophagus. The large concrete enclosure built to contain the Chernobyl nuclear reactor's damaged core.

scattered radiation. A form of secondary radiation in which radiation is scattered in multiple directions during its passage through a substance.

scintillation detector. An assembly that uses phosphor, a photomultiplier tube, and associated electronic circuits to count light emissions produced in the phosphor by ionizing radiation. (2)

scram. The sudden shutting down of a nuclear reactor, usually by rapid insertion of control rods, either automatically or manually by the reactor operator. May also be called a reactor trip. It is an acronym for "safety control rod axe man," the worker assigned to insert the emergency rod on the first reactor (the Chicago Pile) in the United States. (2)

sealed source. Any radioactive material or byproduct encased in a capsule designed to prevent leakage or escape of the material. (2)

secondary radiation. Radiation originating as the result of absorption of other radiation in matter. It may be either electromagnetic or particulate in nature. (2)

secondary system. The steam generator tubes, steam turbine, condenser, and associated pipes, pumps, and heaters used to convert the heat energy of the reactor coolant system into mechanical energy for electrical generation. Most commonly used in reference to pressurized water reactors. (2)

seismic category I. Structures, systems, and components that are designed and built to withstand the maximum potential earthquake stresses for the particular region where a nuclear plant is sited. (2)

self-sustaining chain reaction. Caused when the amount of fissile material present reaches critical mass. *See* Critical mass.

separation of isotopes by laser excitation (SILEX). A technology developed in the 1990s for uranium *isotope separation* to produce *enriched uranium* using *lasers*. Details of the process are classified by the U.S. government.

severe accident. A type of accident that may challenge safety systems at a level much higher than expected. (2)

shallow-dose equivalent (SDE). The external exposure dose equivalent to the skin or an extremity at a tissue depth of 0.007 centimeters (7 mg/cm^2) averaged over an area of 1 square centimeter. (2)

shielding. Any material or obstruction that absorbs radiation and thus tends to protect personnel or materials from the effects of ionizing radiation. (2)

shutdown. A decrease in the rate of fission (and heat production) in a reactor (usually by the insertion of control rods into the core). (2)

shutdown margin. The instantaneous amount of reactivity by which the reactor is subcritical or would be subcritical from its present condition assuming all full-length rod cluster assemblies (shutdown and control) are fully inserted, except for the single rod cluster assembly of highest reactivity worth that is assumed to be fully withdrawn. (2)

sievert (Sv). The international system (SI) unit for dose equivalent, equal to 1 joule/kilogram. 1 Sv = 100 rem. (2)

smart grid. An approach that can be used by utilities to reduce energy use and includes optimizing location of meters and switches so that the system functions more efficiently, not wasting energy. *See* "Global Warming and Fuel Sources."

smart growth. Plans that concentrate development in already urbanized areas and discourage low-density sprawl. *See* "Global Warming and Fuel Sources."

somatic cell. A cell of the body that is not a sperm or egg.

somatic effects of radiation. Effects of radiation limited to the exposed individual, as distinguished from genetic effects, that may also affect subsequent unexposed generations. (2)

source material. Uranium or thorium, or any combination thereof, in any physical or chemical form or ores that contain by weight 1/20 of 1% (0.05%) or more of uranium, thorium, or any combination thereof. Source material does not include special nuclear material. (2)

special form radioactive material. Radioactive material that meets the following criteria: (*a*) is either a single solid piece or is contained in a sealed capsule that can be opened only by destroying the capsule; (*b*) the piece or capsule has at least one dimension not less than 5 mm (0.2 in); and (*c*) it satisfies the requirements of 10 CFR 71.75. A special form encapsulation designed in accordance with the requirements of 10 CFR 71.4 in effect on June 30, 1983 (see 10 CFR pt. 71, revised as of January 1, 1983), and constructed before July 1, 1985, and a special form encapsulation designed in accordance with the requirements of 10 CFR 71.4 in effect on March 31, 1996 (see 10 CFR pt. 71, revised as of January 1, 1983), and constructed before April 1, 1998, may continue to be used. Any other special form encapsulation must meet the specifications of this definition. (2)

special nuclear material. Plutonium, uranium-233, or uranium enriched in the isotopes uranium-233 or uranium-235. (2)

spent (depleted) fuel. Fuel that has been removed from a nuclear reactor because it can no longer sustain power production for economic or other reasons. A recent refinement is between used nuclear fuel, which can be recovered to produce new fuel, and spent fuel which cannot be used for that purpose.

spent fuel pool. An underwater storage and cooling facility for spent fuel elements that have been removed from a reactor. (2)

spent fuel storage cask. All the components and systems associated with the container in which spent fuel or other radioactive materials associated with spent fuel are stored in an independent spent fuel storage installation. Also referred to a simply a cask. (2)

stable isotope. An isotope that does not undergo radioactive decay. (2)

standard review plan. A document that provides guidance to the staff for reviewing a prospective licensee's application to obtain a U.S. Nuclear Regulatory Commission license to construct or operate a nuclear facility or to possess or use nuclear materials. (2)

standard technical specifications. U.S. Nuclear Regulatory Commission staff guidance on model technical specifications for an operating license. (2) *See also* Technical specifications.

startup. An increase in the rate of fission (and heat production) in a reactor (usually by the removal of control rods from the core). (2)

stay time. The period during which personnel may remain in a restricted area in a reactor before accumulating some permissible occupational dose. (2)

steam generator. The heat exchanger used in some reactor designs to transfer heat from the primary (reactor coolant) system to the secondary (steam) system. This design permits heat exchange with little or no contamination of the secondary system equipment. (2)

stochastic effects. Effects that occur by chance, whose probability is proportional to the dose but whose severity is independent of the dose. In the context of radiation protection, the main stochastic effects are cancer and genetic effects. (2)

subcritical mass. An amount of fissionable material insufficient in quantity or configuration to sustain a fission chain reaction. (2)

subcriticality. The condition for decreasing the level of operation of a reactor, in which rate of fission neutron production is less than overall neutron losses. (2)

supercritical reactor. A reactor in which the power level is increasing with time. (2)

supercriticality. The condition for increasing the level of operation of a reactor, in which the rate of fission neutron production exceeds all neutron losses, and the overall neutron population increases. (2)

superheating. The heating of a vapor, particularly steam, to a temperature much higher than the boiling point at the existing pressure. This is done in some power plants to improve efficiency and to reduce water damage to the turbine. (2)

surface-contaminated object (SCO). A solid object that is not itself classed as radioactive material but that has radioactive material distributed on any of its surfaces. (2)

survey meter. Any portable radiation detection instrument especially adapted for inspecting an area or individual to establish the existence and amount of radioactive material present. (2)

tailings. Waste produced by the mining of uranium, which contains radium, which decays to radon, a radioactive gas that will not decay entirely for thousands of years.

technical specifications. Part of a U.S. Nuclear Regulatory Commission license authorizing the operation of a nuclear production or utilization facility. A technical specification establishes requirements for items such as safety limits, limiting safety system settings, limiting control settings, limiting conditions for operation, surveillance requirements, design features, and administrative controls. (2)

terrestrial radiation. The portion of the natural background radiation that is emitted by naturally occurring radioactive materials, such as uranium, thorium, and radon in the earth. (2)

thermal breeder reactor. A breeder reactor in which the fission chain reaction is sustained by thermal neutrons. (2)

thermal power. The total core heat transfer rate to the reactor coolant. (2)

thermal pulse. Heat released in a nuclear explosion, first released as a short, weak pulse and then by a stronger pulse which can last up to 20 seconds.

thermal radiation. Energy that radiates from hot surfaces in the form of electromagnetic waves.

thermal reactor. A reactor in which the fission chain reaction is sustained primarily by thermal neutrons. Most current reactors are thermal reactors. (2)

thermal shield. A layer, or layers, of high-density material located within a reactor pressure vessel or between the vessel and the biological shield to reduce radiation heating in the vessel and the biological shield. (2)

thermalization. The process undergone by high-energy (fast) neutrons as they lose energy by collision. (2)

thermoluminescent dosimeter. A small device used to measure radiation by measuring the amount of visible light emitted from a crystal in the detector when exposed to ionizing radiation. (2)

thermonuclear. An adjective referring to the process in which very high temperatures are used to bring about the fusion of light nuclei, such as those of the hydrogen isotopes deuterium and tritium, with the accompanying liberation of energy. (2)

threshold. The dose or exposure level below which a significant adverse effect is not expected. The lowest dose of a chemical at which a specified measurable effect is observed and below which it is not observed. (4)

total effective dose equivalent (TEDE). The sum of the deep-dose equivalent (for external exposures) and the committed effective dose equivalent (for internal exposures). (2)

transient. A change in the reactor coolant system temperature or pressure due to a change in power output of the reactor. Transients can be caused (*a*) by adding or removing neutron poisons, (*b*) by increasing or decreasing electrical load on the turbine generator, or (*c*) by accident conditions. (2)

transuranic element (TRU). An artificially made, radioactive element that has an atomic number higher than uranium in the periodic table of elements, such as neptunium, plutonium, americium, and others. (2)

transuranic waste (TRUW). Waste that is independent of state or origin and that has been contaminated with alpha emitting transuranic radionuclides possessing half-lives greater than 20 years and in concentrations >100 nCi/g (3.7 MBq/kg) (excluding high-level waste). In the United States it is a byproduct of weapons production and consists of protective gear, tools, residue, debris, and other items contaminated with small amounts of radioactive elements (mainly plutonium).

trip, reactor. A term that is used by pressurized water reactors for a reactor scram. (2) *See* Scram.

tritium. A radioactive isotope of hydrogen with one proton and two neutrons. Because it is chemically identical to natural hydrogen, tritium can easily be taken into the body by any ingestion path. It decays by beta emission and has a radioactive half-life of about 12.5 years. (2)

turbine. A rotary engine made with a series of curved vanes on a rotating shaft, usually turned by water or steam. Turbines are considered the most economical means to turn large electrical generators. (2)

turbine generator (TG). A steam (or water) turbine directly coupled to an electrical generator. The two devices are often referred to as one unit. (2)

type A or type B container. Container types used to transport low-level waste. Type A containers are used to ship most low-level waste; heavier metal engineered casks or type B containers, having been tested under normal and accident conditions, are used to transport higher level waste.

ultraviolet. Electromagnetic radiation of a wavelength between the shortest visible violet and low-energy x-rays. (2)

uncertainty range. Defines an interval within which a numerical result is expected to lie within a specified level of confidence. The interval often used is the 5–95 percentile of the distribution reporting the uncertainty. (2)

underground mine. A mine where coal is produced by tunneling into the earth to the coalbed, which is then mined with underground mining equipment, such as cutting machines and continuous, longwall, and shortwall mining machines. Underground mines are classified according to the type of opening used to reach the coal, such as drift (level tunnel), slope (inclined tunnel), or shaft (vertical tunnel). (3)

unnecessary regulatory burden. Regulatory criteria that go beyond the levels that would be reasonably expected to be imposed on licensees, since regulations apply to conditions that incorporate normal operation and design-basis conditions. (2)

unrestricted area. The area outside the owner-controlled portion of a nuclear facility in which a person must not be exposed to radiation levels in excess of 2 millirems in any 1 hour from external sources. (2)

unstable isotope. A radioactive isotope. (2)

unwanted radioactive material (orphan sources). Sealed sources of radioactive material that meet one or more of the following conditions: (*a*) in an uncontrolled condition that requires removal to protect public safety; (*b*) material for which a responsible party cannot be readily identified; (*c*) controlled material, but the material's continued security cannot be assured; (*d*) in the possession of a person not licensed to possess the material and did not seek to possess the material; or (*e*) in the possession of a state radiological protection for the sole purposing of mitigating a threat caused by one of the above conditions, but for which the state does not have means to provide for the material's appropriate permanent disposition. (2)

uranium. A radioactive element with the symbol U and atomic number 92 and, as found in natural ores, an atomic weight of approximately 238. The two principal natural isotopes are U-235 (0.7% of natural uranium), which is fissile, and U-238 (99.3% of natural uranium), which is fissionable by fast neutrons and is fertile. Natural uranium also includes a minute amount of U-234. (2)

uranium concentrate. A yellow or brown powder obtained by the milling of uranium ore, processing of in situ leach mining solutions, or as a byproduct of phosphoric acid production. (3)

uranium fuel fabrication facility. A facility that (*a*) manufactures reactor fuel containing uranium for any of the following: preparation of fuel materials; formation of fuel materials into desired shapes; application of protective cladding; recovery of scrap material; and storage associated with such operations; or (*b*) conducts research and development activities. (2)

uranium hexafluoride. A compound used in the uranium enrichment process that produces fuel for nuclear reactors or weapons. Referred to as "hex."(3)

uranium hexafluoride production facility. A facility that receives natural uranium in the form of ore concentrate, processes the concentrate, and converts it into uranium hexafluoride (UF6). (2)

vapor. The gaseous form of substances that are normally in liquid or solid form. (2)

very high radiation area. An area accessible to individuals in which radiation levels exceed 500 rad (5 gray) in 1 hour at 1 meter from the source. (2)

viability assessment. A U.S. Department of Energy decision-making process to assess the prospects for geologic disposal of high-level radioactive wastes at Yucca Mountain (NV) based on the design elements of the repository, an assessment of the probable performance of the repository, and an estimate of licensing, construction and operation costs. The viability assessment was required by the Energy and Water Development Appropriations Act, 1997 (P.L. 104-206).

vitrification. A process wherein concentrated process waste from a nuclear site is encapsulated into a glass matrix. *See* "Nuclear Waste Policy."

void. In a nuclear power reactor, an area of lower density in a moderating system (such as steam bubbles in water) that allows more neutron leakage than does the more dense material around it. (2)

void coefficient of reactivity. A rate of change in the reactivity of a water reactor system resulting from a formation of steam bubbles as the power level and temperature increase. (2)

vulnerability analysis (VA). Assessment of elements in the community that are susceptible to damage if hazardous materials are released. (4)

waste, radioactive. Radioactive materials at the end of a useful life cycle or in a product that is no longer useful and should be properly disposed of. (2)

Waste Isolation Pilot Plant (WIPP). A U.S. Department of Energy waste disposal facility located near Carlsbad, NM, for disposal of the nation's defense-related transuranic radioactive waste.

watt. An electrical unit of power. 1 watt = 1 joule/second. It is equal to the power in a circuit in which a current of one ampere flows across a potential difference of one volt. (2)

watt-hour. An electrical energy unit of measure equal to 1 watt of power supplied to, or taken from, an electrical circuit steadily for 1 hour. (2)

weapons of mass destruction. Nuclear, biological, or chemical weapons.

weapons grade material. Contains almost pure (over 90%) plutonium-239, which is created in a reactor specially designed and operated to produce plutonium-239 from uranium—to be distinguished from "reactor grade" plutonium produced as a byproduct in a nuclear power reactor.

weighting factor (WT). Multipliers of the equivalent dose to an organ or tissue used for radiation protection purposes to account for different sensitivities of different organs and tissues to the induction of stochastic effects of radiation. (2)

well-logging. All operations involving the lowering and raising of measuring devices or tools that contain licensed material or are used to detect licensed materials in wells for the purpose of obtaining information about the well or adjacent formations that may be used in oil, gas, mineral, groundwater, or geological exploration. (2)

wheeling service. The movement of electricity from one system to another over transmission facilities. Wheeling service contracts can be established between two or more systems. (2)

whole-body counter. A device used to identify and measure the radioactive material in the body of human beings and animals. It uses heavy shielding to keep out naturally existing background radiation and ultrasensitive radiation detectors and electronic counting equipment. (2)

whole-body exposure. Includes at least the external exposure of head, trunk, arms above the elbow, or legs above the knee. Where a radioisotope is uniformly distributed throughout the body tissues, rather than being concentrated in certain parts, the irradiation can be considered as whole-body exposure. (2)

wind turbine. Wind energy conversion device that produces electricity; typically three blades rotating about a horizontal axis and positioned up-wind of the supporting tower. (3)

wipe sample. A sample made for the purpose of determining the presence of removable radioactive contamination on a surface. It is done by wiping, with slight pressure, a piece of soft filter paper over a surface. It is also known as a "swipe" or "smear" sample. (2)

x-rays. Penetrating electromagnetic radiation (photons) having a wavelength that is much shorter than that of visible light. These rays are usually produced by excitation of the electron field around certain nuclei. In nuclear reactions, it is customary to refer to photons originating in the nucleus as x-rays. (2)

yellowcake. Product of the uranium-mining process; named for the bright yellow compound that resulted from early production methods. The material is a mixture of uranium oxides that can vary in proportion and color from yellow to orange to dark green. Yellowcake is commonly referred to as U_3O_8. This fine powder is packaged in drums and sent to a conversion plant that produces uranium hexafluoride (UF6) as the next step in the manufacture of nuclear fuel. (2)

Sources

1. U.S. Department of Health and Human Services. Radiation Event Medical Management. *Dictionary of radiological terms. www.remm.nlm.gov/dictionary.htm.*
2. U.S. Nuclear Regulatory Commission. *Full text glossary. www.nrc.gov/reading-rm/basic-ref/glossary/full-text.html.*
3. U.S. Department of Energy. Energy Information Administration. *Glossary. www.eia.doe.gov/glossary/index.html.*
4. U.S. Environmental Protection Agency. *Terms of environment: Glossary, abbreviations, and acronyms. www.epa.gov/OCEPAterms.*
5. U.S. Environmental Protection Agency. Software for Environmental Awareness. *Comparative risk analysis. www.epa.gov/seahome/comprisk.html.*
6. U.S. Department of Energy. Waste Isolation Pilot Plant. *Glossary of terms. www.wipp.energy.gov/library/rhwaste/rhsec6.htm.*
7. U.S. Nuclear Regulatory Commission. *Fact sheet: Three Mile Island accident. www.nrc.gov/reading-rm/doc-collections/fact-sheets/3mile-isle.pdf.*
8. U.S. Nuclear Regulatory Commission. *Emergency planning zones. www.nrc.gov/about-nrc/emerg-preparedness/protect-public/planning-zones.html.*
9. AJ Software & Multimedia. AtomicArchive.com. *Glossary. www.atomicarchive.com/Glossary/Glossary1.shtml.*
10. U.S. Environmental Protection Agency. *Radiation protection: A–Z index. www.epa.gov/radiation/atozindex.html.*
11. U.S. Department of Energy. *The Global Nuclear Energy Partnership. www.gnep.energy.gov/gnepProgram.html.*
12. U.S. Environmental Protection Agency. *Green building. www.epa.gov/greenbuilding.*
13. International Atomic Energy Agency. *The "Atoms for Peace" Agency. www.iaea.org/About/index.html.*
14. International Commission on Radiological Protection, home page. *www.icrp.org.*
15. U.S. Nuclear Regulatory Commission. *Low specific activity. www.nrc.gov/reading-rm/basic-ref/glossary/low-specific-activity.html.*
16. Defense Technical Information Center. *Multiple independently targetable reentry vehicle. www.dtic.mil/doctrine/jel/doddict/data/m/03600.html.*
17. Organization for Economic Co-Operation and Development. *The Nuclear Energy Agency. www.nea.fr/html/nea/flyeren.html.*
18. Oak Ridge Institute for Science and Education home page. *orise.orau.gov.*
19. National Cancer Institute. *Dictionary of cancer terms. www.cancer.gov/templates/db_alpha.aspx?expand=P.*
20. U.S. Environmental Protection Agency. Public health and environmental radiation protection

standards for Yucca Mountain, Nevada. *www.epa.gov/radiation/docs/yucca/RIN%202060-an15-final-40-cfr-197amendments.pdf.*

21. U.S. Nuclear Regulatory Commission. *Frequently asked questions about potassium iodide. www.nrc.gov/about-nrc/emerg-preparedness/protect-public/ki-faq.html.*
22. U.S. Nuclear Regulatory Commission. *SECY-06-0185, August 23, 2006. www.nrc.gov/reading-rm/doc-collections/commission/secys/2006/secy2006-0185/2006-0185scy.html.*
23. Organization for Economic Co-Operation and Development. *Radioactive Waste Management Committee. www.nea.fr/html/rwm/rwmc.html.*

Contributors

Experts Interviewed or Consulted

Mark Abkowitz holds an appointment as Professor of Civil and Environmental Engineering at Vanderbilt University and serves as Director of the Vanderbilt Center for Environmental Management Studies. Dr. Abkowitz manages the risks associated with accidents, intentional acts, and natural disasters. He has a special interest in hazardous materials transportation safety and security, and in risk mitigation using advanced information technologies.

James Bresee is currently a Chemical Engineer at Argonne National Laboratory. He has served in many positions related to nuclear energy and waste, including Assistant Director to the Engineering Science and Technology Division at Oak Ridge National Laboratory, Assistant Director for General Energy Development at the U.S. Atomic Energy Commission, Director for the North Carolina Energy Institute, and Director of Repository Coordination Division, Office of Civilian Waste Management, U.S. Department of Energy.

Robert J. Budnitz is on the scientific staff at the University of California's Lawrence Berkeley National Laboratory, where he is Associate Program Leader for nuclear power safety and security and radioactive waste management. Former positions were at the U.S. Department of Energy's Office of Civilian Radioactive Waste Management, Lawrence Livermore National Laboratory, and the U.S. Nuclear Regulatory Commission's Office of Nuclear Regulatory Research, where he was Director. He was President of Future Resources Associates, Inc., for 20 years. He has served on numerous investigative and advisory panels.

Joanna Burger is Distinguished Professor of Biology at Rutgers University. She was a founding member of the Consortium for Risk Evaluation and Stakeholder Participation, serves on the management board, and leads the Ecological Health Center. Her interests include environmental evaluation, ecotoxicology and ecological risk, biomonitoring and indicator development, effects of temperature and contaminants on behavioral development, risks and benefits of fish consumption, stakeholder-driven research, and stakeholder involvement.

Caron Chess, an Associate Professor in Rutgers University's Department of Human Ecology, conducts research on public participation and risk communication. She has authored peer-reviewed publications in academic journals as well as materials that are used widely by government and industry practitioners. She served as President of the Society for Risk Analysis.

James H. Clarke is Professor of the Practice of Civil and Environmental Engineering and Professor of Earth and Environmental Sciences at Vanderbilt University. He is a member of the Nuclear Regulatory Commission Advisory Committee on Nuclear Waste and

Materials. Dr. Clarke received his PhD in theoretical chemistry from Johns Hopkins University.

B. John Garrick's fields of practice are risk assessment and nuclear science and engineering. A founder of the firm PLG, Inc., he was appointed Chairman of the U.S. Nuclear Waste Technical Review Board in 2004 by President George W. Bush and elected to the National Academy of Engineering in 1993. He served as President of the Society for Risk Analysis in 1989–90 and received that society's most prestigious award, the Distinguished Achievement Award, in 1994.

Michael Gochfeld is Professor of Environmental and Occupational Medicine in the Environmental and Occupational Health Sciences Institute at UMDNJ-Robert Wood Johnson Medical School and an original member of the Consortium for Risk Evaluation and Stakeholder Participation. He specializes in occupational medicine, environmental health and ecotoxicology, and risk assessment.

Holly Harrington works in the Office of Public Affairs at the Nuclear Regulatory Commission. Previously, she held public affairs positions at the Federal Emergency Management Agency and the Department of Veterans Affairs and worked as a newspaper reporter. She is an Adjunct Professor at Trinity College, Washington DC. She holds a BA in journalism and an MS in mass communications.

Tom Henry created the *Toledo Blade*'s environment-energy beat and began writing about nuclear power shortly after joining the newspaper in 1993. In 2006, he was awarded a fellowship at Vermont Law School and is consistently ranked by *U.S. News & World Report* as one of the nation's finest writers about environmental law.

Kathryn Higley is a Professor of Nuclear Engineering at Oregon State University and a member of the Consortium for Risk Evaluation and Stakeholder Participation management board. She teaches undergraduate and graduate classes on radioecology, dosimetry, radiation protection, radiochemistry, societal aspects of nuclear technology, and radiation biology. Her fields of interest include environmental transport and fate of radionuclides, radiochemistry, radiation dose assessment, neutron activation analysis, nuclear emergency response, and environmental regulations.

Paul L. Joskow is the Elizabeth and James Killian Professor of Economics and Management at MIT and Director of the MIT Center for Energy and Environmental Policy Research. Dr. Joskow has been on the MIT faculty since 1972 and served as Head of the MIT Department of Economics. He is a Director of National Grid PLC, a Director of TransCanada Corporation, and a Trustee of the Putnam Mutual Funds. He has served on the U.S. Environmental Protection Agency's Acid Rain Advisory Committee and on the Environmental Economics Committee of the agency's Science Advisory Board.

P. Andrew Karam is a board-certified radiation safety professional with over 25 years of experience. He currently is a private consultant and an Adjunct Professor at the Rochester Institute of Technology and serves on committees for the National Council on Radiation Protection and Measurements and for the National Academy of Sciences. Dr. Karam has worked extensively on topics associated with natural radiation on the earth and in space, and on problems associated with radiological terrorism. He is board certified in Health Physics.

David S. Kosson is Professor and Chair of the Department of Civil and Environmental Engineering at Vanderbilt University, where he also has joint appointments as Professor of Chemical Engineering and Professor of Earth and Environmental Sciences. He is Co-principal Investigator (with Charles W. Powers) of the Consortium for Risk Evaluation with Stakeholder Participation. Dr. Kosson's research focuses on containment mass transfer applied to groundwater, soil, sediment, and waste systems, as well as management of nuclear wastes.

Paul Lisowski serves as the Deputy Assistant Secretary for Fuel Cycle Management in the Office of Nuclear Energy, with responsibility for planning and development of advanced fuel cycle facilities and research. He worked at Los Alamos National Laboratory before joining the U.S. Department of Energy.

Paul Meier is Director of the University of Wisconsin Energy Institute. He is a professional engineer with a decade of experience providing energy and environmental consulting to industry, government, and public interest groups. His research focuses on greenhouse gas reduction strategies for the electric industry, for which he has twice received the General Motors Environmental Excellence Award. Dr. Meier is the developer of My Power, a Web-based simulation tool enabling broad public analysis of future energy alternatives.

Frank L. Parker, Distinguished Professor of Environmental and Water Resources Engineering and member of the National Academy of Engineering, is a pioneer in nuclear waste management and environmental protection. Over the past 5 decades, he has served as head of the Radioactive Waste Disposal Research Section of Oak Ridge National Laboratory, head of the Radioactive Waste Disposal Research Program at the International Atomic Energy Agency, senior research fellow of the Beijer Institute of the Royal Swedish Academy of Sciences, and senior research fellow of the International Institute for Applied Systems Analysis (IIASA) in Laxenburg, Austria. Dr. Parker has chaired or been a member of many national and international advisory committees.

Raymond M. Plieness is the Deputy Director for the U.S. Department of Energy (DOE) Office of Legacy Management (LM), Office of Site Operations. He also oversees the LM site transition teams that work with entities transferring sites into LM. Prior to the creation of LM, he served as a project manager, team leader, and the Deputy Manager for DOE's Grand Junction Office, and site transition coordinator for the Rocky Flats, Colorado, Environmental Technology Site.

Michael T. Ryan is a Certified Health Physicist and an independent consultant in radiological sciences and health physics. He works with a variety of private sector and government clients. Dr. Ryan received his PhD from Georgia Institute of Technology, where he was recently inducted into the Academy of Distinguished Alumni. He is a recipient of the Francis Cabot Lowell Distinguished Alumni for Arts and Sciences Award for the University of Massachusetts Lowell.

Buzz Savage is Deputy Assistant Secretary for Fuel Cycle Management in the Office of Nuclear Energy, where he focuses on technology. He has been involved in management of nuclear energy R&D programs at the U.S. Department of Energy as a federal employee and contractor for the past 16 years. Programs in which he has been involved include New Production Reactors, Advanced Light Water Reactors, Generation IV, Advanced Accelerator Applications, and the Advanced Fuel Cycle Initiative. Earlier, he served 20 years in the nuclear navy, operating and maintaining nuclear propulsion plants on nuclear-powered cruisers and aircraft carriers. He is a graduate of the U.S. Naval Academy and holds an MS in nuclear physics from the U.S. Naval Postgraduate School.

Victor Sidel, a physician, is Professor of Social Medicine at Montefiore Medical Center and the Albert Einstein College of Medicine. He has served as President of the American Public Health Association and of Physicians for Social Responsibility and is co-editor of *War and Public Health* (Oxford University Press, 2008).

Paul Slovic is President of Decision Research and a Professor of Psychology at the University of Oregon. He studies human judgment, decision making, and risk perception and has published extensively on these topics. Dr. Slovic received a BA degree from Stanford University and MA and PhD degrees from the University of Michigan.

Michael Stabin is an Associate Professor in the Department of Radiology and Radiological Sciences at Vanderbilt University, in Nashville, Tennessee. He has over 150 publications in the open literature, most in the area of internal dosimetry for nuclear medicine applications, including complete textbooks on health physics and internal dose assessment.

Jane B. Stewart directs the International Environmental Legal Assistance Program at the New York University School of Law. Having served for many years as Senior Staff Attorney in charge of hazardous waste programs for the Natural Resources Defense Council and practiced environmental law at the law firm Paul, Weiss, Rifkind, Wharton & Garrison in New York, she currently advises governmental and nongovernmental organizations in the United States and developing countries on environmental law and policy reform initiatives.

Richard B. Stewart is University Professor and Director of the Center on Environmental and Land Use Law at New York University and Advisory Trustee of Environmental Defense. He is a specialist in environmental, administrative, and regulatory law.

William Szymanski is a Senior Program Manager in the Office of Nuclear Energy, U.S. Department of Energy, with 20 years of experience in the domestic and international nuclear fuel cycle. He has served as a consultant to the International Atomic Energy Agency and frequently speaks at nuclear industry meetings.

Seth Tuler is a Senior Researcher at the Social and Environmental Research Institute. His research interests are focused on the human dimensions of natural resource management and environmental remediation, including public participation and risk communication. He served on the National Academy of Science's Committee on Transportation of Spent Nuclear Fuel and High Level Radioactive Waste and is currently a member of the U.S. Environmental Protection Agency Board of Scientific Advisors' Subcommittee for the National Center for Environmental Research.

Detlof Von Winterfeldt is Director of the University of Southern California Center for Risk and Economic Analysis of Terrorist Events, which is funded by the U.S. Department of Homeland Security. Dr. Von Winterfeldt's research background is in decision and risk analysis applied to environmental, technology, and security problems. He has served on several committees and panels of the National Science Foundation and the National Academies and is currently serving on the National Academy of Sciences Board on Mathematical Sciences and Their Applications.

Reviewers

John F. Ahearne is Director of the Sigma Xi Ethics Program, Adjunct Scholar for Resources for the Future, and a Lecturer in Public Policy at Duke University. A former Chairman of the U.S. Nuclear Regulatory Commission and Deputy Assistant Secretary of Energy, he has chaired and served on numerous national and international panels and committees on nuclear issues. He chaired the National Academy Board on Radioactive Waste Management and co-chaired the Department of Energy Nuclear Energy Advisory Committee, and he is a member of the National Academy of Engineering.

Seth Blumsack is an Assistant Professor in the Department of Energy and Mineral Engineering, Pennsylvania State University. He earned a BA in Mathematics and

Economics from Reed College, and an MS in Economics and a PhD in Engineering and Public Policy from Carnegie Mellon University.

Thomas Cotton, Vice President of JK Research Associates, specializes in radioactive waste management policy analysis and strategic planning in support of the U.S. Department of Energy's high-level radioactive waste program. Before joining JK Research Associates in 1986, he dealt with energy and radioactive waste issues at the Congressional Office of Technology Assessment.

Eric Darois has over 30 years of experience as a Health Physics professional in various technical and managerial positions in nuclear power facilities, decommissioning sites, environmental laboratories, and others. As Principal and Founder of Radiation Safety & Control Services, Inc., he provides radiation protection services that include complex calculations and regulatory compliance support.

Penelope A. Fenner-Crisp, currently an independent consultant, previously held senior scientific and managerial positions at the International Life Sciences Institute and the U.S. Environmental Protection Agency. Her areas of expertise include human health and environmental risk assessment, toxicology, science policy and its integration into regulatory decision-making, and familiarity with environmental regulatory programs and practices.

Keith Florig is Senior Research Engineer in the Department of Engineering and Public Policy at Carnegie Mellon University, where he conducts research in risk analysis and risk communication. His research includes many studies of radiation risk that have been published in *Science, Health Physics,* and other professional journals.

Richard L. Garwin is IBM Fellow Emeritus at the IBM Thomas J. Watson Research Center, Yorktown Heights, New York. He has worked in many areas of physics, technology, and technology and security policy and has published some 600 papers and received 45 U.S. patents. He is a winner of the National Medal of Science.

Don Hopey has covered the environment beat for the *Pittsburgh Post-Gazette* since 1992 and is co-author of *Exploring the Appalachian Trail: Hikes in the Mid-Atlantic States* (Stackpole, 1998), a book highlighting the trail's social and natural history. He is a board member of the Society of Environmental Journalists and teaches an environmental issues and policy class at the University of Pittsburgh.

Thomas H. Isaacs is the Director of the Office of Policy, Planning, and Special Studies at Lawrence Livermore National Laboratories. He has held a variety of positions within the U.S. Department of Energy (DOE), including Executive Director of DOE's Advisory Committee on External Regulation of DOE Nuclear Safety, Director of Strategic Planning and International Programs, and Deputy Director of the Office of Geologic Repositories for the department's radioactive waste program. He has served on the National Research Council's Committees on Principles and Operational Strategies for Staged Repository Systems, and Building a Long-Term Environmental Quality Research and Development Program in DOE.

David C. Kocher is a Senior Scientist at SENES Oak Ridge, Inc., Center for Risk Analysis. He has more than 30 years of experience in environmental health physics, assessments of radioactive waste disposal, and assessments of human health risks from exposure to radiation.

Sandra Quinn is with the Graduate School of Public Health, University of Pittsburgh, where she teaches risk communication. She was the Principal Investigator on a study of risk communication during the anthrax attack and is currently the guest editor for an issue on emergency risk communication and pandemic influenza for Health Promotion Practice.

Milton Russell is a Senior Fellow of the Institute for a Secure and Sustainable Environment (ISSE) and a Professor Emeritus of Economics at the University of Tennessee. His current research focuses on analysis and policy direction for managing the environmental legacy of U.S. Department of Energy facilities. Before coming to Tennessee in 1987, Russell served as an Assistant Administrator of the U.S. Environmental Protection Agency, directing its policy, planning, regulatory development, and evaluation functions. He accepted the position as founding Director of the Joint Institute for Energy and Environment (now merged into ISSE) in late 1992; prior to that he held a joint appointment with the University of Tennessee at Knoxville Economics Department and the university's Energy, Environment, and Resource Center (now merged into ISSE).

Niel Wald has specialized in radiation hematology and cytogenetics and radiation injury management, serving in the U.S. Air Force, the Atomic Bomb Casualty Commission (Hiroshima), and Oak Ridge National Laboratory and, since 1958, the University of Pittsburgh's Graduate School of Public Health, where he now is Emeritus Professor of Environmental and Occupational Health.

J. Samuel Walker is Historian of the U.S. Nuclear Regulatory Commission. He is the author of *Three Mile Island: A Nuclear Crisis in Historical Perspective* (University of California Press, 2004) and other books on the history of nuclear energy and regulation.

Chris Whipple is a Principal in Environ International's Emeryville, California, office. His expertise is with the assessment of risks associated with radioactive wastes, mercury, and hazardous air pollutants. He has served on committees of the National Academy of Sciences, the U.S. Environmental Protection Agency, and the National Council on Radiation Protection and Measurement.

Authors

Michael R. Greenberg is Professor and Associate Dean of the Faculty of the Edward J. Bloustein School of Planning and Public Policy, Rutgers University. Dr. Greenberg has written extensively about environmental and public health policies, is editor-in-chief of *Risk Analysis: An International Journal,* and associate editor for environmental health for the *American Journal of Public Health.*

Karen W. Lowrie is a Research and Program Associate at the National Center for Neighborhood and Brownfields Redevelopment at the Bloustein School of Planning and Public Policy, Rutgers University. Dr. Lowrie conducts research and outreach in the areas of neighborhood redevelopment, environmental planning, and media analysis. She is also Managing Editor of *Risk Analysis: An International Journal.*

Henry J. Mayer is Executive Director of the National Center for Neighborhood and Brownfields Redevelopment at Rutgers University, where his research focuses on the environmental, infrastructure, capital financing, and land use issues associated with the redevelopment of distressed cities. He also works with the Consortium for Risk Evaluation with Stakeholder Participation (CRESP) to examine U.S. Department of Energy's environmental management efforts and has published numerous articles, reports, and book chapters on these topics.

Bernadette M. West is an Associate Professor in the Health Systems and Policy Department at the University of Medicine and Dentistry of New Jersey–School of Public Health. She also serves as Assistant Dean of the Stratford/Camden Campus and Assistant Dean for Community Health.

Index